# Essays in Biochemistry

**Other recent titles in the Essays in Biochemistry series:**

Essays in Biochemistry volume 37: Regulation of Gene Expression
*edited by* K.E. Chapman and S. J. Higgins
2001 ISBN 1 85578 138 7

Essays in Biochemistry volume 36: Molecular Trafficking
*edited by* P. Bernstein
2000 ISBN 1 85578 131 X

Essays in Biochemistry volume 35: Molecular Motors
*edited by* G. Banting and S.J. Higgins
2000 ISBN 1 85578 103 4

Essays in Biochemistry volume 34: Metalloproteins
*edited by* D.P. Ballou
1999 ISBN 1 85578 106 9

Essays in Biochemistry volume 33: Molecular Biology of the Brain
*edited by* S.J. Higgins
1998 ISBN 1 85578 086 0

Essays in Biochemistry volume 32: Cell Signalling
*edited by* D. Bowles
1997 ISBN 1 85578 071 2

volume 38 2002

# Essays in Biochemistry

## Proteases in Biology and Medicine

Edited by N.M. Hooper

Portland Press

*Essays in Biochemistry* is published by Portland Press Ltd
on behalf of the Biochemical Society

**Portland Press**
**59 Portland Place**
**London W1B 1QW, U.K.**
**Fax: 020 7323 1136;**
**e-mail: editorial@portlandpress.com**
**www.portlandpress.com**



Although, at the time of going to press, the information contained in this publication is believed to be correct, neither the authors nor the editors nor the publisher assumes any responsibility for any errors or omissions herein contained. Opinions expressed in this book are those of the authors and are not necessarily held by the Biochemical Society, the editors or the publisher.

All profits made from the sale of this publication are returned to the Biochemical Society for the promotion of the molecular life sciences.

**British Library Cataloguing-in-Publication Data**
A catalogue record for this book is available from the British Library

ISBN 1 85578 147 6
ISSN 0071 1365

Typeset by Portland Press Ltd
Printed in Great Britain by Information Press Ltd, Oxford

# Contents

## 3 Matrix metalloproteinases in cancer

*Yoshifumi Itoh and Hideaki Nagase*



## 4 Proteolytic processing of the amyloid-β protein precursor of Alzheimer's disease

*Janelle Nunan and David H. Small*

## 5 The ubiquitin–proteasome pathway of intracellular proteolysis

*Fergus J. Doherty, Simon Dawson and R. John Mayer*



## 6 Methionine aminopeptidases and angiogenesis

*Ralph A. Bradshaw and Elizabeth Yi*



## 7 Precursor convertases in the secretory pathway, cytosol and extracellular milieu

*Nabil G. Seidah and Annik Prat*

## 8 Proteases in blood clotting

*Peter N. Walsh and Syed S. Ahmad*



## 9 Anatomy and pathology of HIV-1 peptidase

*Ben M. Dunn*



## 10 Inhibition of peptidases in the control of blood pressure

*Eiji Kubota, Rachael G. Dean, Leanne C. Balding and Louise M. Burrell*

## 11 Shedding of membrane proteins by ADAM family proteases

*Marcia L. Moss and Millard H. Lambert*



## 12 Regulated intramembrane proteolysis: from the endoplasmic reticulum to the nucleus

*Robert B. Rawson*

## 13 Protease-activated receptors: the role of cell-surface proteolysis in signalling

*Graeme S. Cottrell, Anne-Marie Coelho and Nigel W. Bunnett*



## 14 Mining proteases in the genome databases

*David Coates*

# Preface

Proteases are being recognized more and more as important players in a wide range of biological processes; for example, the cell cycle, blood clotting, angiogenesis, apoptosis, cell differentiation and growth, cell motility, lipid metabolism, antigen presentation and cell-fate determination are all dependent upon the controlled action of proteases. When the activity of proteases is not regulated appropriately, disease processes can result, as is seen in Alzheimer's disease, cancer metastasis and tumour progression, inflammation, pain, atherosclerosis and haemophilia. Thus, inhibition of proteases is being seen as a potential therapeutic strategy for these and other disorders. Probably the most successful examples of protease inhibitors in medicine today are the inhibitors of angiotensin-converting enzyme that are widely used to treat hypertension and congestive heart failure, and the HIV protease inhibitors that are at the forefront of the battle against this virus.

This volume covers many of these topics and highlights the role of proteases in biology and medicine. The first chapter introduces the reader to the basic terminology in the field of protease research and the classification systems used, and this is followed by a chapter by Guy Salvesen, who expertly introduces the caspases, the executioners of cell death (apoptosis). Yoshifumi Itoh and Hideaki Nagase then provide an excellent account of the large family of matrix metalloproteinases and their role in cancer metastasis, while Janelle Nunan and David Small outline the role of proteases (secretases) in the processing of the Alzheimer's amyloid precursor protein, and the potential sites of therapeutic intervention in this disease. In the subsequent chapter, John Mayer and colleagues describe the ubiquitin pathway of intracellular proteolysis and the central role played by 'the big mean proteolytic machine', the proteasome, in this process. The next three chapters cover the roles that proteases play in angiogenesis (Ralph Bradshaw and Elizabeth Yi), the processing of protein precursors (Nabil Seidah and Annik Prat), and the blood clotting cascade (Peter Walsh and Syed Ahmad). The two chapters that follow highlight the potential use of protease inhibitors as therapeutic agents, with a description of the HIV peptidase by Ben Dunn, and the use of angiotensin-converting enzyme inhibitors by Louise Burrell and colleagues. The latter chapter also introduces other peptidases as possible future drug targets in hypertension and heart disease. The next three chapters introduce topics that have only recently come to the fore. Marcia Moss and Millard Lambert provide an overview of membrane-protein shedding and the role of the ADAMs family of proteases in this process, Rob Rawson introduces the fascinating area of regulated intramembrane proteolysis, while Nigel Bunnett and colleagues review the surprising role that protease-activated receptors have to play in cell signalling. The final chapter, by David Coates, covers the important place that bioinformatics and genome database screening has, and will continue to have, in the identification of novel proteases.

With any volume of this nature, it is inevitable that many proteases have been omitted and to those whose favourite protease is not covered, I apologize. I would like to thank all the authors for their scholarly contributions and for keeping them within strict length limits. My hope is that this volume will provide a taster to the exciting field of protease research for senior undergraduates, junior postgraduates and even seasoned researchers. Finally, my thanks are owing to Sophie Dilley and the other staff at Portland Press Ltd for their hard work in the production of this book.

**Nigel M. Hooper**
*Leeds, 2002*

# Authors

**Nigel Hooper** is Professor of Biochemistry in the School of Biochemistry and Molecular Biology at the University of Leeds. He received his B.Sc. in Biochemistry in 1981 and his Ph.D. in 1984, both from the University of Leeds. His Ph.D. study on the 'Metabolism of neuropeptides by cell-surface peptidases' stimulated his interest in proteases. After a 2-year post-doctoral position at Leeds, he joined the academic staff as a lecturer in 1989. At present, he co-leads the Proteolysis Research Group with Tony Turner, and among other topics, he continues to study the structure and function of several cell-surface proteases, with a particular interest in their mode of attachment to the membrane.

**Guy Salvesen** is Professor and Director of the Program in Apoptosis and Cell Death Research at the Burnham Institute, La Jolla, CA, and Adjunct Professor of Molecular Pathology at the University of California, San Diego, CA. He received his Ph.D. after studying the regulation of proteolysis under the supervision of Alan Barrett at Cambridge University. After post-doctoral training on inflammatory proteases with James Travis at the University of Georgia, he moved to Duke University, Durham, NC, as Assistant Research Professor, where he started his work on caspases. He moved to the Burnham Institute in 1996.

**Yoshifumi Itoh** is a Senior Lecturer of Matrix Biology at the Kennedy Institute of Rheumatology, Faculty of Medicine, Imperial College of Science, Technology and Medicine, London. He received his B.Sc. in Pharmacy in 1989 and his M.Sc. in Clinical Pharmacy in 1991 from Tokyo University of Pharmacy and Life Science. He then became a Research Associate working on matrix metalloproteinases in Hideaki Nagase's laboratory at the University of Kansas Medical Center, Kansas City, KS, and was awarded a Ph.D. in Pharmaceutical Science in 1996 by Tokyo University of Pharmacy and Life Science. In 1997, he moved back to Tokyo and became a *Joshu* (Lecturer) of the Department of Cancer Cell Research in Motoharu Seiki's laboratory at the Institute of Medical Science, University of Tokyo. He assumed his present position in 2001 and his research interests involve investigating the role of pericellular proteolysis in cell migration. **Hideaki Nagase** is Professor of Matrix Biology at the Kennedy Institute of Rheumatology Division, Imperial College of Science, Technology and Medicine, London. He received his B.Sc. in Pharmacy from Tokyo University of Pharmacy and Life Science in 1971, an M.Sc. in Physiological Chemistry from the Science University of Tokyo in 1973 and his Ph.D. in Biochemistry from the University of Miami in 1977. He was an Assistant Professor of Medicine at Robert Wood Johnson Medical

School, University of Medicine and Dentistry of New Jersey, and a Professor of Biochemistry and Molecular Biology at the University of Kansas Medical Center. He assumed his present position in 1999 and his research interests involve investigating the structure, function and pathophysiological roles of matrix metalloproteinases and their inhibitors.

**Janelle Nunan** is a Ph.D. student in the Laboratory of Molecular Neurobiology at the University of Melbourne, Australia. Her doctoral studies focus on amyloid-β protein precursor secretases. **David H. Small** is Head of the Laboratory of Molecular Neurobiology at the University of Melbourne, Australia. After postdoctoral work in the early 1980s at Massachusetts Institute of Technology and Flinders University, Adelaide, he moved to the University of Melbourne. His work has focused on the role of proteases in the trafficking of proteins in the central nervous system, and on the biochemistry and cell biology of cholinesterases.

**Fergus Doherty** is a Lecturer in Biochemistry in the School of Biomedical Sciences at Nottingham University, and his research interests include the role of the ubiquitin-like protein UCRP (ubiquitin cross-reactive protein) in cells of the immune system and the human endometrium. **Simon Dawson** is a Lecturer in Molecular Biology in the School of Biomedical Sciences at Nottingham University, and during the last few years he has been involved in employing the yeast two-hybrid technique to study the molecular interactions of proteasomal subunits. **John Mayer** is Professor of Molecular Cell Biology and head of the Intracellular Proteolysis Laboratory in the School of Biomedical Sciences at Nottingham University. Professor Mayer's current research interests include the molecular interactions of proteasome subunits with each other and with non-proteasomal proteins.

**Ralph A. Bradshaw** is a Professor in the Department of Physiology and Biophysics at the University of California at Irvine. He holds degrees from Colby College, Waterville, ME, and Duke University, and has been a researcher/faculty member at Indiana University, University of Washington, Seattle, WA, and Washington University in St. Louis, MO. He is a member of several learned societies, including the American Society for Biochemistry and Molecular Biology and the Protein Society, and has served as an Associate Editor of the *Journal of Biological Chemistry* and *Protein Science*. He is also the founding editor of *Molecular and Cellular Proteomics*. He has a long-term interest in protein structure and the function of proteases, and in polypeptide growth factors and their receptors. **Elizabeth Yi** is a staff research associate in the Department of Physiology and Biophysics at the University of California at Irvine. She holds a B.S. degree from Long Beach State University, CA.

**Nabil Seidah** received his Ph.D. degree in biophysics and physical chemistry from Georgetown University, Washington, DC, in 1973. After a brief post-doctoral training, he joined Michel Chrétien at the Clinical Research Institute of Montreal in 1974, where he has been ever since. From 1974 to

1989, his research dealt with the characterization of various polypeptide hormones, including β-endorphin and atrial natriuretic factor, as well as the definition of their biosynthetic pathways. In 1987, he spent a sabbatical year at the Pasteur Institute, Paris, to study renin processing and activation. On his return to Montreal, he cloned and characterized the proprotein convertases PC1 and PC2. This led him to identify the other members of the family, such as PC4, PC7 and SKI-1. He is now continuing his work on the structure–function of the convertases and is actively exploring their implication in proliferative and neurodegenerative diseases. **Annik Prat** received her Ph.D. in Cellular and Molecular Biology from the Pierre and Marie Curie University, Paris in 1988. After a 2.5-years of post-doctoral training in Patrick Linder's laboratory at the Biozentrum in Basel, Switzerland, she joined Paul Cohen's group in Paris in 1990. Between 1990 and 1997, she cloned and studied a newly characterized enzyme, the *N*-arginine dibasic converter (NRDc). After 1 year in Guy Boileau's laboratory at the University of Montreal, she joined Nabil Seidah's team where she is pursuing her work on NRDc.

**Peter N. Walsh**, M.D., Ph.D., is a physician/scientist in the Departments of Medicine and Biochemistry at Temple University School of Medicine, and at the Sol Sherry Thrombosis Research Center, Philadelphia, PA. He received his undergraduate degree at Amherst College, Amherst, MA, in 1957, an M.D. degree from Washington University School of Medicine in 1961, and the D.Phil. from the University of Oxford in 1972. His research interests include the interactions of coagulation proteins with platelets, and the biochemistry of blood coagulation proteins, especially factors XI, IX, VIII and X. **Syed S. Ahmad**, M.D., Ph.D., is a Research Professor of Biochemistry and the Sol Sherry Thrombosis Research Center at Temple University School of Medicine. He received his Ph.D. from the University of Karachi, Pakistan, in 1979, and his M.D. degree from University of Ciudad. Juarez, Mexico, in 1983. His research interests include the structural and functional relationships of coagulation proteins and their interaction with platelets.

**Ben Dunn** is Distinguished Professor of Biochemistry and Molecular Biology at the University of Florida College of Medicine. He received a B.S. degree in Chemistry from the University of Delaware in 1967, and a Ph.D. from the University of California, Santa Barbara, in 1971. From 1971 to 1974, he was a post-doctoral fellow and staff fellow at the National Institutes of Health, Bethesda, MD. His research interests are in the structure and function of enzymes, particularly the peptidases. He has contributed to studies of the substrate and inhibitor specificity of the aspartic peptidase family of enzymes, including those from HIV-1 and feline immunodeficiency virus, the malarial parasite *Plasmodium falciparum*, fungi and humans.

**Eiji Kubota** is a Visiting Fellow in the Department of Medicine, University of Melbourne. He undertook his medical training in Japan and qualified as a specialist in Internal Medicine in 1995. His research interests include the role of the renin–angiotensin system and bradykinin in hypertension and renal disease, as well as the pharmacology of vasopeptidase inhibitors and calcium antagonists. He will be returning to Japan in 2002 to work at the Shizuoka Red Cross Hospital Department of Medicine, Shizuoka-ken. **Rachael G. Dean** is a post-doctoral fellow in Department of Medicine, University of Melbourne. She obtained her Ph.D. in 1997 and her research interests include the roles of vaso-active peptides in the kidney and heart and of growth factors involved in cardiac remodelling. **Leanne Balding** undertook her medical training at Monash University, Melbourne, and her cardiology training at the Royal Melbourne Hospital. She currently works as a clinical cardiologist at the Royal Melbourne and Freemasons' Hospitals and is completing her Doctor of Medicine in the Department of Medicine, University of Melbourne. Her research has focused on the use of vasopeptidase inhibitors and vasopressin receptor antagonists in heart failure. She is the recipient of the Viola Edith Reid Bequest Scholarship, a Cardiac Society of Australia and New Zealand Research Grant, an Austin Research Medical Foundation Grant and a Pfizer CVL Research Grant. **Louise M. Burrell** is an Associate Professor of Medicine in the Department of Medicine, University of Melbourne, and a General Physician/Endocrinologist. She obtained a British–Australian Heart Foundation Fellowship in 1991 and now runs a Cardiovascular Endocrinology Group that investigates cardiac and renal aspects of hypertension, heart failure and diabetes.

**Mill Lambert** is a computational chemist at GlaxoSmithKline in Research Triangle Park, NC. He received his B.A. in Physics at the University of Virginia in 1982, and Ph.D. in Physics at Cornell University, NY, in 1988. During his doctoral and post-doctoral work with Harold Scheraga, he developed the Molecular Viewing Program for protein modelling, molecular docking and structure-based drug design. Dr Lambert has worked on a number of drug discovery projects at Glaxo, GlaxoWellcome and GlaxoSmithKline, including those for phospholipase A2, collagenase, tumour necrosis factor-α-converting enzyme, and peroxisome-proliferator-activated receptor-α, -γ and -δ. **Marcia L. Moss** received her B.S. in chemistry from the University of Michigan, Ann Arbor, in 1982 and her Ph.D. in biochemistry from the University of Wisconsin, Madison, in 1989. After two post-doctoral positions, she began working at Glaxo Inc. During her time at Glaxo, she headed the work that purified TNF-α-converting enzyme (TACE). She later led a project that studied matrix metalloproteinases in cancer. She retired from Glaxo Wellcome owing to complications from multiple sclerosis. At present, she works part-time at Cognosci Inc., a company that is researching anti-inflammatory treatments for neurological and other diseases.

**Rob Rawson** studied botany and history at the University of California, Berkeley, and graduated in 1982. After a few years in the computer business, he earned a Master's degree from the California State University, Hayward, in 1987. He received a Ph.D. from the University of Texas Southwestern Graduate

School of Biomedical Science in 1993. He then did a postdoctoral fellowship in the laboratory of Michael Brown and Joseph Goldstein in the Department of Molecular Genetics at University of Texas Southwestern Medical Center. In 1999, he joined that Department as a faculty member, where he continues research into the control of lipid metabolism in vertebrates and insects.

**Nigel Bunnett** is a Professor of Surgery and Physiology at the University of California, San Francisco and is the Director of the Gastrointestinal Research Center. He obtained a Ph.D. degree at Cambridge and completed post-doctoral training at University of California, Los Angeles. **Graeme Cottrell** and **Anne-Marie Coelho** are post-doctoral fellows in Nigel Bunnett's research laboratory at the University of California, San Francisco. Graeme received a Ph.D. degree from the University of Leeds, where he studied aminopeptidases with Nigel Hooper and Tony Turner in the School of Biochemistry and Molecular Biology. Anne-Marie obtained her Ph.D. degree in the Neuro-Gastroenterology Unit at the University of Toulouse, where she studied the role of proteases and their receptors in hyperalgesia and inflammation with Dr Bueno.

**David Coates** originally trained as a plant molecular biologist at the John Innes Institute, Norwich, and Purdue University, West Lafayette, IN, before moving to Leeds via Oxford to pursue his interests in the evolution of duplications and small gene families using the classic molecular model systems of *Caenorhabditis elegans* and *Drosophila melanogaster*, as well as a broad set of genomes that include the plant RNA viruses. The enormous changes in the amount of information available to biologists is now driving a deep ambition to understand how to program a computer!

# Abbreviations

| | |
|---|---|
| Aβ | amyloid-β protein |
| ACE | angiotensin-converting enzyme |
| AD | Alzheimer's disease |
| ADAM | a disintegrin and metalloproteinase |
| Ang | angiotensin |
| ANP | atrial natriuretic peptide |
| APC | anaphase-promoting complex |
| APP | amyloid-β protein precursor |
| ATF | activating transcription factor |
| BACE | β-site APP-cleaving enzyme |
| BDNF | brain-derived neurotrophic factor |
| BIR | baculovirus IAP repeat |
| CA | capsid protein |
| Cbz | benzyloxycarbonyl |
| CARD | caspase recruitment domain |
| Cdk | cyclin-dependent kinase |
| CGRP | calcitonin gene related peptide |
| CHIP | C-terminal Hsp70-interacting protein |
| DED | death effector domain |
| DUB | de-ubiquitylating enzyme |
| E1 | ubiquitin-activating enzyme |
| E2 | ubiquitin-conjugating enzyme |
| E3 | ubiquitin–protein ligase |
| ECM | extracellular matrix |
| EGF | epidermal growth factor |
| EIAV | equine infectious anemic virus |
| eIF | eukaryotic initiation factor |
| EPR-1 | effector cell protease receptor-1 |
| ER | endoplasmic reticulum |
| FAD | familial Alzheimer's disease |
| FGF | fibroblast growth factor |
| FIV | feline immunodeficiency virus |
| FLIP | Flice-like inhibitory protein |
| GP-C | 76-kDa precursor glycoprotein of LAV |
| GPCR | G-protein-coupled receptor |
| GPI | glycosylphosphatidylinositol |
| HB-EGF | heparin-binding EGF-like growth factor |

| | |
|---|---|
| HK | high molecular mass kininogen |
| HMM | hidden Markov model |
| Hpx domain | hemopexin-like domain |
| IAP | inhibitor of apoptosis protein |
| IκB | inhibitor of NF-κB |
| IDE | insulin-degrading enzyme |
| IGF | insulin-like growth factor |
| IGFBP | insulin-like growth factor-binding protein |
| LAV | Lassa virus |
| LCMV | lymphocytic choriomeningitis virus |
| MAP kinase | mitogen-activated protein kinase |
| MetAP | methionine aminopeptidase |
| MHC I | Major Histocompatability Complex Class I |
| MMP | matrix metalloproteinase |
| MT-MMP | membrane-type matrix metalloproteinase |
| NC | nucleocapsid protein |
| NEP | neutral endopeptidase |
| NFκB | nuclear factor κB |
| NK1R | neurokinin$_1$ receptor |
| NRDc | *N*-arginine dibasic convertase |
| PACE | paired basic amino acid converting enzyme |
| PAI | plasminogen activator inhibitor |
| PAR | protease-activated receptor |
| PC | proprotein convertase |
| α1-PDX | α1-antitrypsin Portland |
| PK | prekallikrein |
| PKC | protein kinase C |
| PN-2 | protease nexin 2 |
| PS | presenilin |
| RA | rheumatoid arthritis |
| Rip | regulated intramembrane proteolysis |
| RSV | Rous sarcoma virus |
| RUP | regulated ubiquitin/proteasome-dependent processing |
| sAPPα | soluble N-terminal fragment of APP |
| SCAP | SREBP cleavage activating protein |
| SCF | Skip–cullin–F-box |
| SERPIN | serine protease inhibitor |
| SET domain | serine-, glutamine- and threonine-rich domain |
| SG | secretory granule |
| SIV | simian immunodeficiency virus |
| SKI-1 | subtilisin/kexin-like isozyme 1 |
| SP | substance P |

| | |
|---|---|
| S1P | site-1 protease |
| S2P | site-2 protease |
| SREBP | sterol regulatory element binding protein |
| STAT | signal transducer and activator of transcription |
| SUMO | small ubiquitin-like modifier |
| SVMP | snake venom metalloprotease |
| TACE | tumour necrosis factor-α-converting enzyme |
| TFPI | tissue factor pathway inhibitor |
| TGF | transforming growth factor |
| TGN | *trans* Golgi network |
| TIMP | tissue inhibitor of metalloproteinases |
| TNF | tumour necrosis factor |
| t-PA | tissue plasminogen activator |
| TRAF | tumour necrosis factor-receptor-associated factor |
| UBP | ubiquitin-specific protease |
| VEGF | vascular endothelial cell growth factor |
| XIAP | X-linked IAP |
| ZPI | protein Z-dependent protease inhibitor |

# 1

# Proteases: a primer

## Nigel M. Hooper[1]

*Proteolysis Research Group, School of Biochemistry and Molecular Biology, University of Leeds, Leeds LS2 9JT, U.K.*

### Abstract

A protease can be defined as an enzyme that hydrolyses peptide bonds. Proteases can be divided into endopeptidases, which cleave internal peptide bonds in substrates, and exopeptidases, which cleave the terminal peptide bonds. Exopeptidases can be further subdivided into aminopeptidases and carboxypeptidases. The Schechter and Berger nomenclature provides a model for describing the interactions between the peptide substrate and the active site of a protease. Proteases can also be classified as aspartic proteases, cysteine proteases, metalloproteases, serine proteases and threonine proteases, depending on the nature of the active site. Different inhibitors can be used experimentally to distinguish between these classes of protease. The MEROPs database groups proteases into families on the basis of similarities in sequence and structure. Protease activity can be regulated *in vivo* by endogenous inhibitors, by the activation of zymogens and by altering the rate of their synthesis and degradation.

### Introduction

For many researchers, proteases are often considered to be unwanted biological pests. Many do their utmost to inactivate proteases in order to prevent the breakdown of their particular protein of interest. For others, proteases are tools to be used to selectively destroy or chop up a protein prior to its further analysis, e.g. the digestion of a protein with trypsin prior to sequencing the smaller tryptic fragments. However, more and more

[1]*E-mail: n.m.hooper@leeds.ac.uk*

researchers arc recognizing that proteases are often key players in a wide range of biological processes; for example, in regulating the cell cycle, cell growth and differentiation, antigen processing and angiogenesis. In addition, it is becoming apparent that the aberrant functioning of certain proteases may be involved in several disease states, including Alzheimer's disease, in cancer metastasis and in inflammation. An understanding of the role played by proteases in these processes may provide the opportunity for therapeutic intervention, and inhibitors of certain proteases have already proved to be effective therapeutic agents in hypertension and heart failure, some forms of cancer, and against certain viruses. The aim of this volume is to highlight some of the more recent developments in this area and to provide an insight into the future of protease research. When one considers that almost 2% of the human genome codes for proteases [1], it is clear that there is a lot still to be learned. The remainder of this chapter is devoted to a brief introduction to the basic terminology used in protease biology.

## Definition of proteases

A protease is defined as an enzyme that hydrolyses one or more peptide bonds (Figure 1) in a protein or peptide [2]. Thus, proteases can, potentially, degrade anything containing a peptide bond, from a dipeptide up to a large protein containing thousands of amino acids. However, many proteases have a preference for protein substrates, while others will only cleave short peptides or even just dipeptides. As these enzymes hydrolyse peptide bonds, some have argued that they all should be termed 'peptidases', and that the term 'protease' be restricted to those peptidases that hydrolyse proteins. Other commonly found terms in the literature include 'proteinase' and 'proteolytic enzyme'.

## Cleavage-site specificity

The terminology used in describing the cleavage-site specificity of proteases is based on a model proposed by Schechter and Berger [3]. In this model, the catalytic site is considered to be flanked on one or both sides by specificity subsites, each of which is able to accommodate the side chain of a single amino acid residue (Figure 2). By convention, the substrate amino-acid residues are called P (for peptide) and the subsites on the protease that interact with the substrate are called S (for subsite). The subsites are numbered outwards from the catalytic site, S1, S2, S3, etc. towards the N-terminus of the substrate, and S1′, S2′, S3′, etc. towards the C-terminus (Figure 2). The side chains of the amino-acid residues in the substrate that these sites accommodate are numbered P1, P2, etc. and P1′, P2′, etc., outwards from the scissile peptide bond (see Figure 2). The residues are usually not numbered beyond P6 on either side of the scissile bond. Different proteases have different requirements for subsite interactions to determine the specificity of cleavage. For example, the S1 subsite of trypsin has a marked preference for the binding of basic

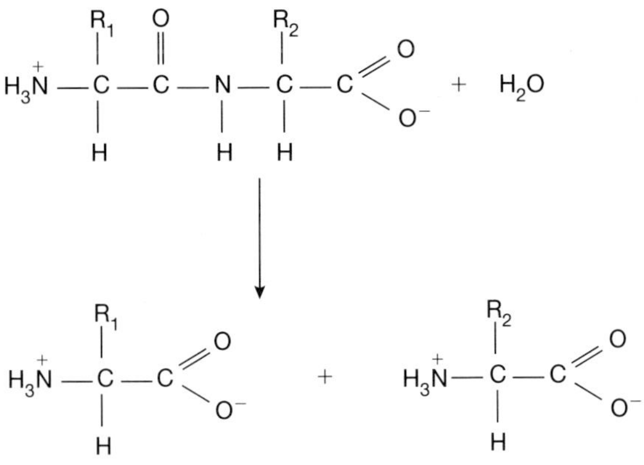

$$H_3\overset{+}{N}-CH(R_1)-CO-NH-CH(R_2)-COO^- + H_2O \rightarrow H_3\overset{+}{N}-CH(R_1)-COO^- + H_3\overset{+}{N}-CH(R_2)-COO^-$$

**Figure 1. Peptide bond hydrolysis by a protease**

amino acid residues (arginine and lysine), while interactions with several of the subsites further away from the scissile bond are critical for substrate binding to renin, the protease involved in the renin–angiotensin system (see Chapter 10), and to the caspases, proteases involved in apoptosis (see Chapter 2).

## Classification of proteases

Why classify proteases? First, classification aids researchers and students in understanding the terminology in this large, and often confusing, field of research. Secondly, the grouping together of enzymes in families on the basis of sequence and structural information aids in the elucidation of common catalytic, biosynthetic processing and regulatory mechanisms. Finally, such classification is invaluable in elucidating the function of newly identified proteases. This is particularly relevant in the context of proteases that are

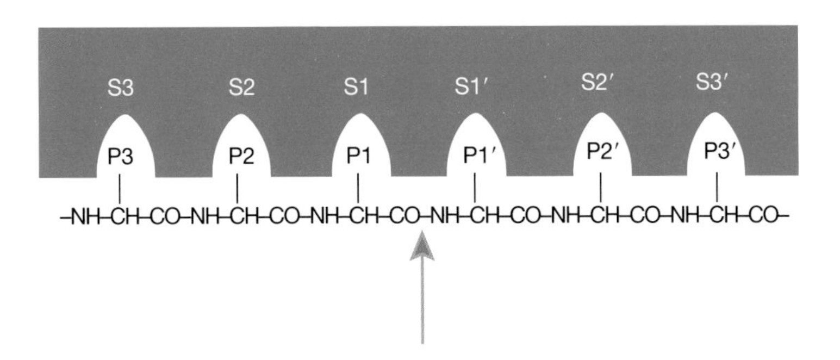


**Figure 2. The Schechter and Berger [3] nomenclature for binding of a peptide substrate to a protease**

The protease is represented as the blue shaded area. P1, P1′, etc. are the side chains of the six amino acids surrounding the peptide bond to be cleaved (indicated by the arrow) in the substrate. S1, S1′, etc. are the corresponding subsites on the protease.

initially identified on the basis of sequence similarity from screening genome databases [1], rather than in the more traditional way of isolating an activity that cleaves a particular protein or peptide substrate, followed by its purification and experimental characterization. Mining of genome databases for novel proteases is dealt with in more detail in Chapter 14.

Proteases can be classified on the basis of the position within a peptide of the peptide bond that is cleaved. Thus, endopeptidases cleave internal peptide bonds, while exopeptidases cleave the terminal bonds (Figure 3). Exopeptidases can be further subdivided into aminopeptidases or carboxypeptidases, depending on whether they cleave the N-terminal or C-terminal peptide bond respectively (Figure 3). Proteases are also classified on the basis of the catalytic mechanism, that is, the nature of the amino acid residue or cofactor at the active site that is involved in the hydrolytic reaction. Thus, aspartic proteases, such as the HIV protease (see Chapter 9) and renin (see Chapter 10), have a critical aspartate residue that is involved in catalysis. Metalloproteases have a bivalent metal ion, usually zinc but sometimes cobalt, iron or manganese, at the active site. Examples of metalloproteases include the matrix metalloproteases (see Chapter 3), methionine aminopeptidases (see Chapter 6), angiotensin-converting enzyme and neprilysin (see Chapter 10), and the ADAMs (a disintegrin and metalloproteinase domain) family of proteases (see Chapter 11). In the aspartic and metalloproteases, the nucleophile that attacks the peptide bond of the substrate is an activated water molecule, whereas in the other protease groups the nucleophile is part of an amino acid at the catalytic site of the protease.

Those proteases in which the nucleophile is the sulphydryl group of a cysteine residue are termed cysteine proteases, typified by the caspases that are involved in programmed cell death (see Chapter 2). In serine proteases the catalytic mechanism depends upon the hydroxy group of a serine residue acting as the nucleophile that attacks the peptide bond. Examples of serine proteases include chymotrypsin and trypsin (digestive enzymes of the intestine), and the proteases involved in the blood clotting cascade (see Chapter 8). In a small number of proteases the catalytic mechanism depends on the hydroxy group of a threonine residue, the so-called threonine proteases. These are exemplified by

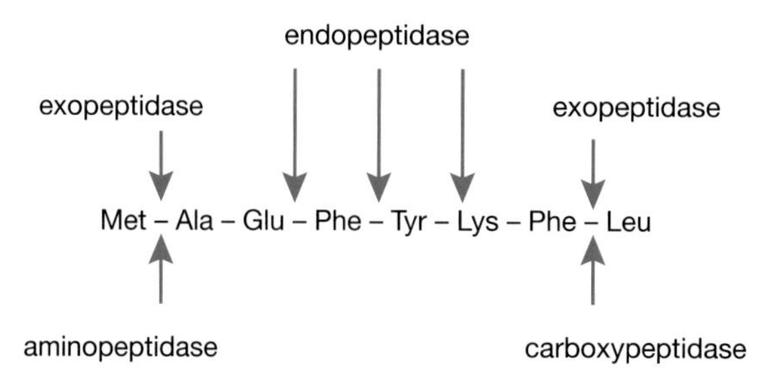


**Figure 3. Cleavage site specificity of proteases**

the catalytic subunit of the proteasome (see Chapter 5). At present, 3% of the proteases in the human genome have been identified as aspartic proteases, 23% as cysteine proteases, 36% as metalloproteases and 32% as serine proteases [1].

Over the past decade, Alan Barrett and his colleagues in Cambridge, U.K., have developed a more detailed classification system for proteases, the MEROPS database [4–6]. This is available online (www.merops.ac.uk) or in hard copy as the *Handbook of Proteolytic Enzymes* [7]. In the MEROPS database, proteases are classified by structural similarities in the parts of the molecules that are responsible for their enzymic activity. They are grouped into families on the basis of amino-acid sequence homology, and the families are assembled into clans based on evidence, usually similarities in tertiary structure, that they share a common ancestry (Figure 4). This classification forms a framework around which a wealth of supplementary information about the proteases is organized, including images of three-dimensional structures, amino-acid sequence alignments, comments on biomedical relevance, and literature references. A set of online searches provides access to information about the location of proteases on human chromosomes and their substrate specificity. As the MEROPS database is updated regularly, it provides an extremely valuable resource for protease researchers (see also Chapter 14).

## Inhibition of proteases

The four major classes of proteases (aspartic, cysteine, metallo and serine) can be distinguished experimentally using class-specific inhibitors (Table 1). For example, chelators such as EDTA or 1,10-phenanthroline remove the critical metal ion from the catalytic site of metalloproteases, thereby inactivating them.

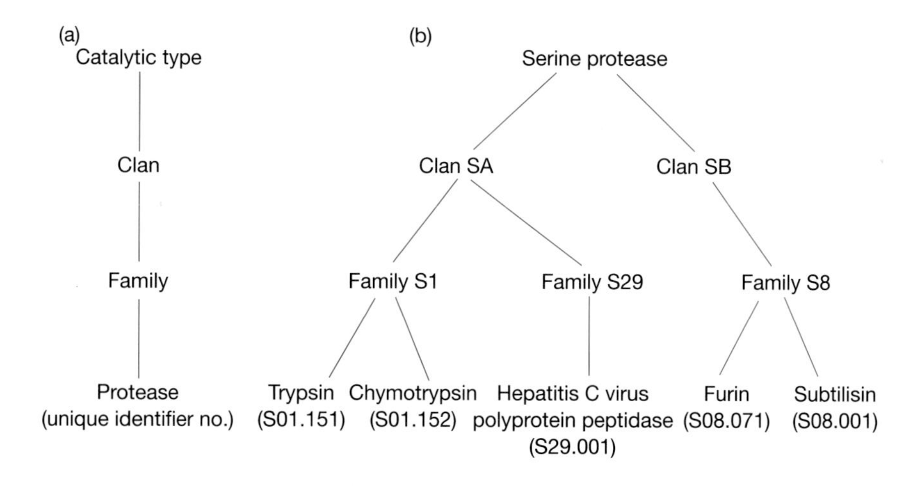


**Figure 4. Overview of the MEROPS protease classification system for proteases**
(a) Summary showing the relationship between catalytic type (aspartic, cysteine, metallo or serine), clan, family and individual protease within the MEROPS database. (b) Schematic showing the relationship of five serine proteases within the MEROPS database. For more information see [7].

On the other hand, di-isopropyl fluorophosphate binds irreversibly to the serine residue at the catalytic site of serine proteases, thus permanently inactivating the enzyme. However, not every protease is susceptible to inhibition by one of these more general class-specific inhibitors. For example, the recently discovered aspartic protease involved in the β-secretase cleavage of the Alzheimer's disease amyloid precursor protein is not inhibited by pepstatin [8] (see Chapter 4).

These class-specific inhibitors (Table 1) can be used experimentally to identify the catalytic type that a particular protease belongs to, and are particularly useful in the absence of amino-acid sequence information. In addition, knowledge of the different protease classes and their respective inhibitors is often useful in designing a strategy to block the breakdown of a particular protein in a sample by the inhibition of either as many proteases as possible or a discrete subgroup of proteases. To this end, inhibitor cocktails, consisting of mixtures of the compounds in Table 1, are available from a number of commercial suppliers.

## Regulation of protease activity

In addition to the inhibition of proteases *in vitro*, there are several ways in which the activities of proteases can be regulated *in vivo*. These include inhibition by endogenous inhibitors, which are often proteins themselves, such as the antitrypsin inhibitor that binds to prematurely activated trypsin in the pancreas, the serine protease inhibitors and Kunitz-type protease inhibitors that inhibit, amongst others, some of the serine proteases involved in blood clotting (see Chapter 8), the tissue inhibitors of metalloproteases that inhibit the matrix metalloproteases (see Chapter 3), and the inhibitor of apoptosis proteins that inhibits the caspases (see Chapter 2).

Protease activity can also be regulated in other ways. The rate of synthesis and/or the rate of degradation will determine the amount of a particular protease present at any one time. Such mechanisms are often used to restrict the expression and activity of a protease to a particular tissue or stage of development. Many proteases are first synthesized in an inactive pro-form, often termed a zymogen, which is itself proteolytically cleaved to the active protease, e.g. trypsinogen is activated to form the digestive enzyme trypsin. Other examples include the matrix metalloproteases (see Chapter 3), the ADAMs family (see Chapter 11) and the serine proteases involved in blood clotting (see Chapter 8). The latter case is an excellent example of a series of zymogen activations finely regulating a biological process. Several zymogens are activated by the removal of their prodomain by serine proteases of the furin/pro-hormone convertase family that cleave at pairs of dibasic residues (see Chapter 7).

**Table 1. Class-specific protease inhibitors**

The mode of inhibition is indicated as either reversible (R) or irreversible (I). More information on individual inhibitors can be found in [9] or the supplier's catalogues.

| Class of protease inhibited | Inhibitor | Mode | Effective concentration | Comments |
|---|---|---|---|---|
| Acidic | Pepstatin A | R | 1 μM | |
| Metallo | EDTA | R | 1–10 mM | |
| | 1,10-Phenanthroline | R | 1–10 mM | Particularly effective against zinc metalloproteases. |
| | Bestatin | R | 1–10 μM | Mainly selective for aminopeptidases. |
| | Phosphoramidon | R | 1–10 μM | Inhibits thermolysin- and neprilysin-like proteases. |
| Cysteine | *trans*-Epoxysuccinyl-L-leucylamido-(4-guanidino)butane (E-64) | I | 1–10 μM | |
| | Iodoacetamide | I | 10–100 μM | Can react with non-active-site cysteine residues. |
| | Leupeptin | R | 10–100 μM | Also inhibits some serine proteases. |
| Serine | Aprotinin | R | 2–10 μg/ml | Is itself a protein. |
| | 4-(2-Aminoethyl)benzenesulphonyl fluoride (AEBSF) | I | 0.1–1 mM | More stable in aqueous solution than PMSF. |
| | Di-isopropyl fluorophosphate (DIPF) | I | 0.1 mM | Extremely toxic. Half-life in aqueous solution for 1h at pH 7.5. |
| | Phenylmethylsulphonylfluoride (PMSF) | I | 0.1–1 mM | Half-life in aqueous solution for 1h at pH 7.5. |

## Summary

- *A protease is an enzyme that hydrolyses peptide bonds.*
- *The Schechter and Berger nomenclature [3] provides a model for describing the interactions between the peptide substrate and the active site of a protease.*
- *Proteases can be subdivided into endopeptidases and exopeptidases, depending on the position in the substrate of the peptide bond that is hydrolysed.*
- *Proteases are classified depending on the nature of the active site into aspartic proteases, cysteine proteases, metalloproteases and serine proteases.*
- *Different inhibitors can be used experimentally to distinguish between these four classes of protease.*
- *The MEROPS database groups proteases into families on the basis of similarities in sequence and structure.*
- *Protease activity can be regulated in vivo by endogenous inhibitors, activation of zymogens, and by altering the rate of their synthesis and degradation.*

## References

1. Southan, C. (2001) A genomic perspective on human proteases. *FEBS Lett.* **498**, 214–218
2. Smith, A.D., Datta, S.P., Howard Smith, G., Campbell, P.N., Bentley, R. & McKenzie, H.A. (1997) *Oxford Dictionary of Biochemistry and Molecular Biology*, Oxford University Press, Oxford
3. Schechter, I. & Berger, A. (1967) On the size of the active site in proteases. I. Papain. *Biochem. Biophys. Res. Commun.* **27**, 157–162
4. Rawlings, N.D. & Barrett, A.J. (1993) Evolutionary families of peptidases. *Biochem. J.* **290**, 205–218
5. Rawlings, N.D. & Barrett, A.J. (1999) MEROPS: the peptidase database. *Nucleic Acids Res.* **27**, 1–7
6. Barrett, A.J., Rawlings, N.D. & O'Brien, E.A. (2001) The MEROPS database as a protease information system. *J. Struct. Biol.* **134**, 95–102
7. Barrett, A.J., Rawlings, N.D. & Woessner, J.F. (1998) *Handbook of Proteolytic Enzymes*, Academic Press, San Diego, CA
8. Vassar, R., Bennett, B.D., Babu-Khan, S., Kahn, S., Mendiaz, E.A., Denis, P., Teplow, D.B., Ross, S., Amarante, P., Leoloff, R. et al. (1999) β-Secretase cleavage of Alzheimer's amyloid precursor protein by the transmembrane aspartic protease BACE. *Science* **286**, 735–741
9. Beynon, R.J. & Bond, J.S. (1989) *Proteolytic Enzymes: A Practical Approach*, IRL Press, Oxford

# 2

# Caspases and apoptosis

## Guy S. Salvesen[1]

*Program in Apoptosis and Cell Death Research, The Burnham Institute, La Jolla, CA 92037, U.S.A.*

### Abstract

The ability of metazoan cells to undergo programmed cell death is vital to both the precise development and long-term survival of the mature adult. Cell deaths that result from engagement of this programme end in apoptosis, the ordered dismantling of the cell that results in its 'silent' demise, in which packaged cell fragments are removed by phagocytosis. This co-ordinated demise is mediated by members of a family of cysteine proteases known as caspases, whose activation follows characteristic apoptotic stimuli, and whose substrates include many proteins, the limited cleavage of which causes the characteristic morphology of apoptosis. In vertebrates, a subset of caspases has evolved to participate in the activation of pro-inflammatory cytokines, and thus members of the caspase family participate in one of two very distinct intracellular signalling pathways.

### Introduction: caspases and their relatives

The peptidases that are members of the group known as clan CD (see Chapter 1) contain several families of related cysteine proteases, which are characterized by their folding pattern or, when this is not available, by secondary structure predictions and the location of active site residues [1]. These peptidases include the bacterial clostripain and gingipains, plant and animal legumains, the anaphase-associated protease separase, and the caspases. A high degree of preference for distinct substrate side-chains at P1 (the residue on the N-terminal side of the scissile peptide bond; for an explanation see Chapter 1, Figure 2) is a

[1]*E-mail: gsalvesen@burnham.org*

characteristic of this clan. This is most clearly demonstrated by the stringent preference for aspartate in the S1 pocket (the region of the protease that interacts with the P1 side-chain) of the caspases. Hence, the origin of the name for the family, which is a contraction of cysteine-dependent aspartate-specific protease. Mammals contain two biologically distinct caspase sub-families: one of these participates in the processing of pro-inflammatory cytokines, while the other is required to elicit and execute the apoptotic response during programmed cell death (Figure 1). Confirmation of the important roles of the caspases in either the inflammatory cytokine response or in apoptosis comes from gene ablation experiments in mice. Animals ablated in caspase 1 are deficient in cytokine processing, but do not have an overt apoptotic phenotype. In contrast, the phenotypes of other knockouts are very severe, evidently anti-apoptotic and include early embryonic lethality (caspase 8), perinatal lethality (caspases 3 and 9) and relatively mild defects in the process of normal oocyte ablation (caspase 2) (reviewed in [2]). At present, caspase 14 may be the odd one out, as it is involved in keratinocyte differentiation [3].

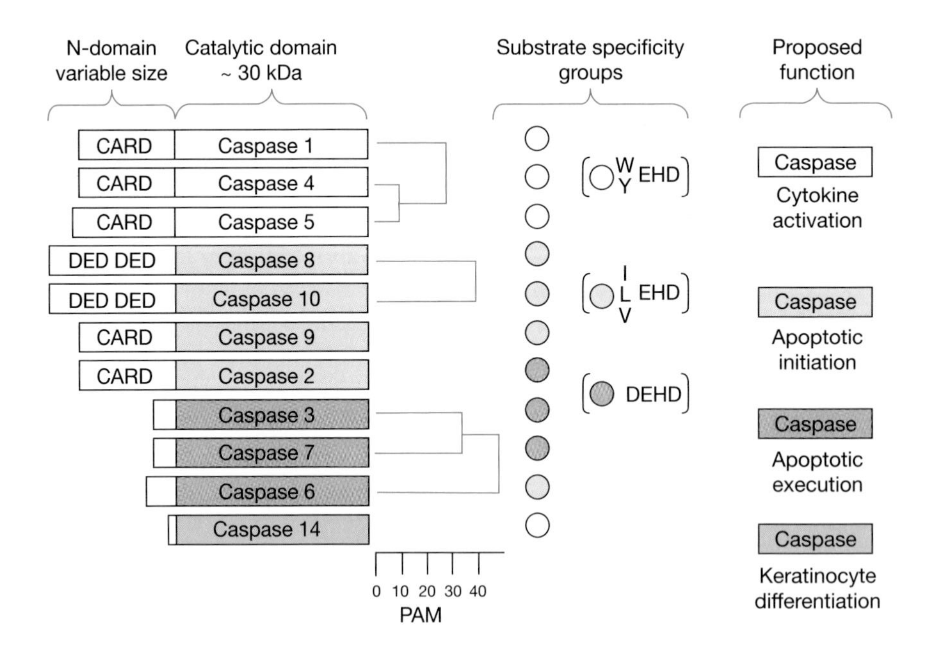


**Figure 1. Human caspase groupings**

Caspases can be grouped together based on sequence homology, which implies evolutionary relatedness, or by substrate specificity in the P4–P1 positions, or by presumed physiological roles based on a number of biological and genetic criteria. The substrate specificity and functional groupings do not always match, as exemplified by caspases 2 and 6, which seem to be out of place because of the expected position of caspase 2 as an initiator caspase, and caspase 6 as an executioner. PAM, point accepted mutations.

## Programmed cell death

The adult human body eliminates in the region of $10^{11}$ cells daily (mostly by apoptosis) to counter the body's proliferative requirements. Programmed cell death has been shown to play a crucial role in all models of metazoan development, from *Caenorhabditis elegans* to the mouse. Many cells that appear during development simply do not exist in the adult animal. The origin of programmed cell death is obscure, but it is apparent that at some stage during the evolution of multicellular organisms, the requirement to delete excess cells became important in establishing the optimal pattern of a functional adult. The ability to target the destruction of specific cells was a vital development in the progress towards complex higher animals, and essentially the same system is used to achieve selectivity of the immune response and maintain cell number in adults. The system has even been co-opted to allow selective killing of virally infected and transformed cells, and it relies on caspases both as initiators of the apoptotic pathway and as executioners (Figure 2). It is this division of labour in the initiation and execution phases that enables cells to respond to external and internal stimuli so as to make the cell death decision, a process that has been fine-tuned through evolution to generate distinct functions within the caspase family.

The N-terminal portion (sometimes called the 'pro-domain') of caspases is not conserved throughout the family and the lengths of this region vary considerably (Figure 1). In contrast, the catalytic core of caspases is well conserved, especially around the catalytic dyad formed by $His^{237}$ and $Cys^{285}$. Other highly conserved residues include those that form the S1 pocket and the oxyanion hole. Cytokine activators (caspases 1, 4 and 5) show greater sequence similarity than any other groups. Other distinct similarity groupings include the initiator caspases 8 and 10 that play a similar role in receptor-stimulated caspase activation, and the executioner caspases 3 and 7 that play a combined role in cleaving the many protein substrates that are the hallmark of apoptosis. Interestingly, the initiator caspases 8 and 10, and the executioner caspases 3 and 7 may have emerged from their respective common ancestors at a stage between bony fishes and frogs.

## Caspase organization and mechanism

### Non-catalytic domains

Two types of well recognized domains can be found at the N-terminal region of several caspases (Figure 1): CARD (caspase recruitment domain) [4] and DED (death effector domain) [5]. Both domains are distantly related, and are implicated in homophilic interactions with other proteins that allow the respective caspases to be recruited to protein complexes for activation. Activation of caspases that contain CARDs and DEDs normally leads to cleavage of these domains and the release of the active catalytic portion (caspase 9 is an exception to this rule).

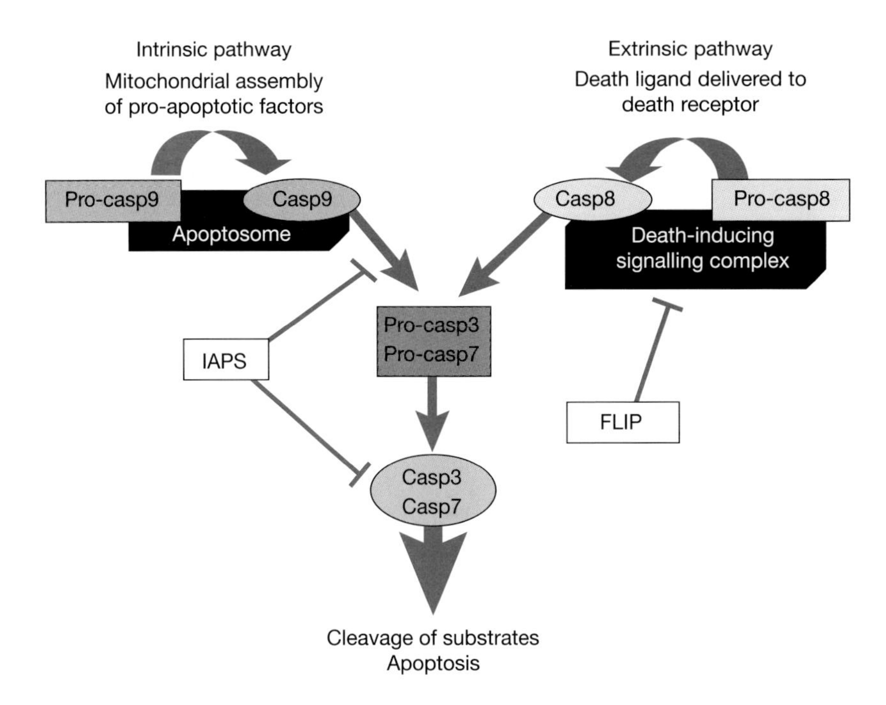


**Figure 2. Division of labour in the intrinsic and extrinsic apoptotic pathways**
The intrinsic pathway responds primarily to cellular stress (ionizing radiation, cytotoxic drugs, etc.), as well as to some neurodevelopmental cues, with the mitochondrion acting as an important integrator. Activation of the apical protease caspase 9 occurs when it is driven into a catalytic conformation at the protein complex known as the apoptosome [31]. The extrinsic apoptosis pathway is triggered through the extracellular ligation of death receptors by their cognate ligands, which results in activation of caspase 8 via the protein complex known as the 'death-inducing signalling complex' [20]. Both pathways activate the common executioner proteases caspases 3 and 7. Natural inhibitors, such as FLIP and IAPs, affect different points on the pathways.

Caspases 3, 6, and 7 have short N-peptides (23–28 residues) that are removed during activation, and less information is available concerning the function of these peptides. With the exception of the caspase 9 CARD, the structures of the N-terminal domains of caspases have not been determined, and at present there is no caspase structure that includes both the N-terminal domain and catalytic domain.

## Catalytic domains

All caspases are synthesized as inactive precursors (zymogens) and show a similar domain organization (Figure 1). The active catalytic units (approx. 30 kDa) are well conserved, are usually composed of large (approx. 20 kDa) and small (approx. 10 kDa) catalytic subunits, and contain all of the enzymic machinery. The crystallographic structures of caspases 1, 3, 7, 8 (reviewed in [6,7]) and caspase 9 [8] have been determined and they all show the same basic folding pattern. The units that comprise each catalytic domain are $\alpha\beta\alpha$ sandwiches, folded into a compact cylinder with six-stranded $\beta$-sheets in the centre, surrounded by five helices. Despite the presence of extra strands in some structures, the overall arrangement is kept intact. Dimerization of the catalytic units appears essential for activity, and each catalytic unit is related by two-fold symmetry. It is often said that the catalytic sites are found at opposite ends of the two units within the dimer, but this is not strictly true, since the angle they subtend is approx. 110°, which places them on the same face of the molecule (Figure 3).

### Active site

A close-up view of the caspase 3 active site (Figure 4a) shows the enzyme with the inhibitor acetyl-Asp-Glu-Val-Asp-CHO bound to the active cysteine residue; the equivalent view of the R-gingipain active site is shown for comparison. The catalytic dyad $His^{237}$ and $Cys^{285}$ is found in the large subunit (amino acid numbering used here is that of caspase 1) and the S1 subsite is formed by side chains from both large and small subunits. Subsites S2–S4 are predominantly set by the small subunit. It is, therefore, the variation in amino acid sequence of the small subunit that is responsible for the variable extended substrate specificity of caspases. The main chain of the substrate forms an antiparallel β-strand with the main chain of enzyme residues 339–341. Interestingly, this mode of substrate alignment is very different to that achieved by other cysteine protease families, but is shared by the chymotrypsin family of serine proteases.

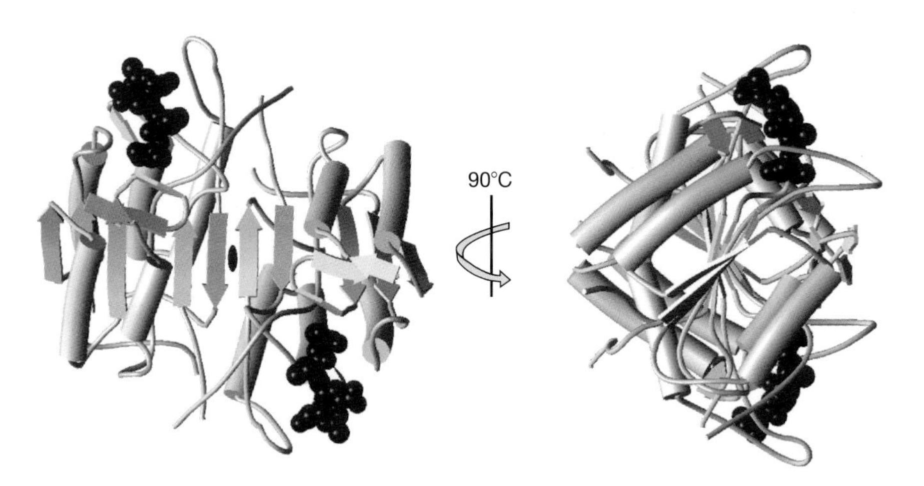


**Figure 3. Caspase 3**
The active molecule is an obligate dimer of catalytic domains, each one containing an apparently equal and independent active site. Each catalytic domain contains a large and a small subunit, with the small subunits providing most of the dimer interface (see also Figure 4). The domains are related by 2-fold symmetry about the *Z*-axis (indicated by the ellipse in the view on the left). The relative placement of the active sites, represented by a tetrapeptide inhibitor in black space-filling form, is more readily apparent in the right view, which has been rotated 90° about the *Y*-axis.

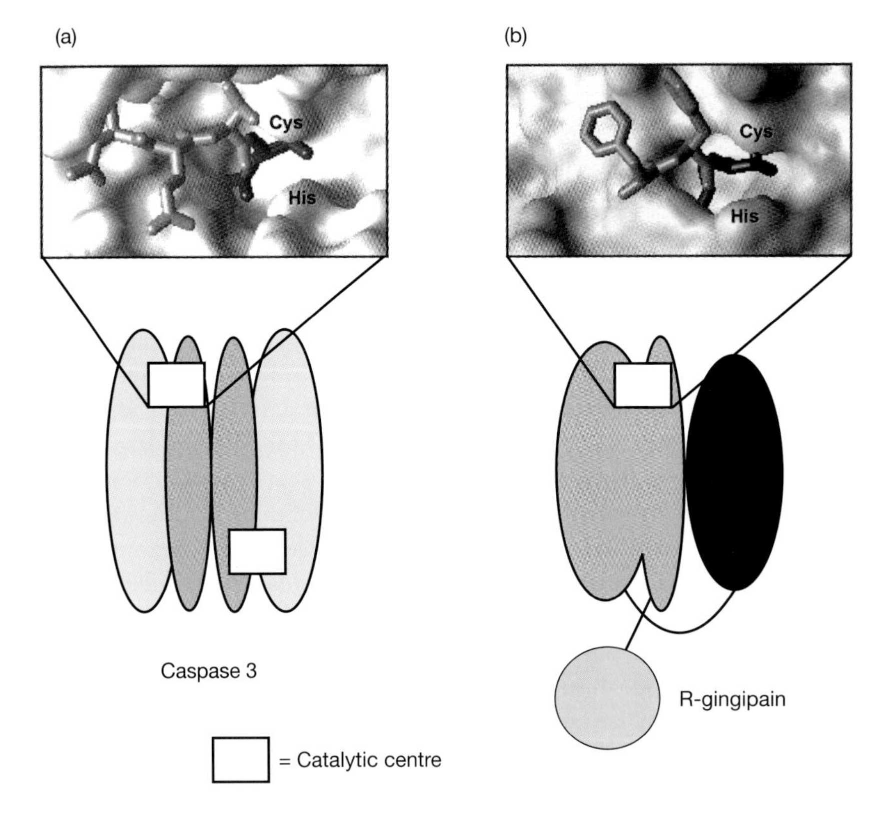


**Figure 4. Caspase 3, its distant homologue R-gingipain and their active sites**
(a) The location of the active site (catalytic centre) of both enzymes is shown. The small subunits of each caspase 3 catalytic domain are depicted in light grey, and the large subunits are in light blue. (b) Unlike caspase 3, R-gingipain is a single-chain protein with its N-terminus in the black domain, and its C-terminus in the blue domain. The N-terminal domain simulates the dimeric partner found in all active caspases, and probably serves to stabilize the catalytic domain. The insets show the active-site regions in standard protease orientation, with the substrate (inhibitor) running from left to right through the catalytic cleft. Caspase 3 has the inhibitor acetyl-Asp-Glu-Val-Asp-CHO bound and R-gingipain has Phe-Phe-Arg-$CH_2Cl$ bound. The respective P1 residues (Asp or Arg) are coloured black, and the locations of the catalytic Cys and His residue surfaces are indicated.

## Catalytic mechanism and substrate specificity

Caspases have a stringent specificity for an aspartate residue in the S1 substrate pocket. The S1 pocket is formed by co-operation of at least three residues ($Arg^{179}$, $Gln^{283}$ and $Arg^{341}$) that form hydrogen bonds with the carboxylate group of the P1 aspartate residue. This tightly controlled environment prevents other amino acids such as glutamate from occupying the S1 pocket; indeed, the selectivity ($k_{cat}/K_m$) for aspartate over glutamate reaches as high as 20000-fold for caspase 3. Caspases display maximal activity at neutral pH (6.8–7.2), in reducing conditions and at the ionic strength of the cytosol [9,10]. The scissile peptide bond of the substrate lies between the catalytic cysteine and histidine side-chains, which is very unusual, since other protease families have their catalytic nucleophile and general base residue on the same side of the scissile bond. This suggests that His[237] does not participate directly in the generation of the catalytic cysteine nucleophile, but rather is important for the protonation of the α-amine of the leaving group and the generation of a nucleophilic water molecule required for deacylation. These observations suggest to us that caspases utilize a catalytic dyad and not a triad. Significantly, a similar arrangement of the catalytic residues is found in R-gingipain (Figure 4b).

Though caspases share a common P1 preference for aspartate, they generally also share a slight preference for histidine at the S2 site, glutamate at the S3 site, and small uncharged residues at S1′ (the site that interacts with the amino acid on the C-terminal side of the scissile bond; for an explanation see Chapter 1, Figure 2) [11,12]. The main specificity determinant that distinguishes the functional caspase groups from each other is the nature of the S4 pocket (Figure 1). These preferences account, in part, for caspase specificity on natural substrates. For example, interleukin 1β is processed at a Tyr-Val-His-Asp site by caspase 1 [13]; caspases 8 and 9 process the zymogens of executioner caspases 3 and 7 at Ile-Glu-Thr-Asp and Ile-Gln-Ala-Asp respectively; and caspases 3 and 7 process the execution substrates such as poly(ADP-ribose) polymerase, which is involved in DNA repair [14], caspase-activated DNase, which regulates DNA laddering [15], and the 70-kDa protein component of the U1 small nuclear ribonucleoprotein, which is involved in mRNA splicing [16], at Asp-Xaa-Xaa-Asp sites.

Using two-dimensional gel electrophoresis, it has been shown that a limited set of proteins is cleaved during apoptosis (approx. 200) [17,18]. Approximately 100 of them have been identified, but only a few have been directly demonstrated to be implicated in apoptosis. The vast majority can be cleaved either as part of the general breakdown of the cell or during ‘post-mortem’ proteolysis. The vast majority of activity has been attributed to caspase 3; there are two reasons for this. First, at a concentration of approx. 200 nM, caspase 3 is the most abundant caspase in cells. Secondly, it has a $k_{cat}/K_m$ for an Asp-Xaa-Xaa-Asp fluorogenic substrate that is at least 6-fold higher than that of its closest relative, caspase 7 [11], which is present at a 5-fold lower concentration [19]. This constitutes a major obstacle to the *in vivo* study of caspase functions, since the activity of other caspases is frequently overpowered by caspase 3 activity.

## Caspase activation

As is the case for almost all proteolytic enzymes, caspases are restrained in an inactive form (the zymogen) until activation by proteolysis (see Chapter 1). The dogma for pro-caspase activation says that they are activated by proteolysis between the large and small subunits. Therefore, a perplexing question posed by the dogma is: how are the initiator caspases 8 and 9 activated if there are no proteases ‘above’ them in the pathway? Indeed, this is a common feature of many proteolytic pathways (coagulation and fibrinolysis for example; see Chapter 8), and not just for apoptosis. This may be partly explained by the induced proximity hypothesis (reviewed in [20]), which proposed that initiator caspases are recruited to protein complexes whose assembly forces a locally high concentration of caspase zymogens. This clustering of zymogens, which possess a small amount of intrinsic enzymic activity, would allow for processing in *trans*, and activation of the first protease in the cascade. However, recent data on the three-dimensional structure of caspase 9 offer another explanation for the conversion of its zymogen to the active form [8].

At *in vivo* concentrations, pro-caspase 9 contains the equivalent of a single domain, but this domain is in an inactive conformation. The substrate binding cleft is dislocated and the catalytic residues are distorted. Induced proximity forced within the Apaf-1 (apoptosis protease-activating factor) apoptosome may lead to dimerization, resulting in an ordering of the active site with development of catalytic competence. This model offers a slightly different explanation than the induced proximity hypothesis, since no proteolytic processing of pro-caspase 9 would be required to drive activation.

Interestingly, although the driving forces are different, the zymogen of the executioner caspase 7 is maintained in a similar inactive conformation to the caspase 9 zymogen [21,22]. The primary difference is that pro-caspase 7, unlike pro-caspase 9, is already a dimer and the driving force for zymogen activation is proteolysis. This releases the N-termini of the small subunits and C-termini of the large subunits, thereby allowing them to form the four-stranded loop bundles that are required to order the catalytic site (Figure 3).

The structures of the caspase 7 zymogen seem to rule out previously proposed models in which the two subunits of the active caspase molecule arise from different subunits of each zymogen by a domain swapping mechanism. Rather, the two catalytic domains in the zymogen are associated by simple dimerization, no domain swapping takes place, and the driving force for activation is the large movement of the inter-chain connector that helps to align the substrate-binding region and catalytic machinery.

## Caspase inhibition

There are two types of endogenous caspase inhibitors in humans. One of these, FLIP (Flice-like inhibitory protein; Flice is an old name for caspase 8), prevents activation of the extrinsic pathway by blocking caspase 8 recruitment and activation (reviewed in [23]). It is not specifically a caspase inhibitor, but a competitor of the activation process. True caspase inhibitors are currently restricted to members of the IAP (inhibitor of apoptosis protein) family [24].

The IAPs are broadly distributed. As their name indicates, the first IAPs to be discovered are capable of selectively blocking apoptosis; they were identified initially in baculoviruses (reviewed in [25]), while eight distinct IAPs have been identified in humans. X-linked IAP (XIAP) (which is the human family paradigm) has been found by several research groups to be a potent, but restricted, inhibitor that targets caspases 3, 7 and 9 (reviewed in [24]). Similarly, evidence implicates human cellular (c)IAPs 1 and 2, melanome (ML)-IAP, *Drosophila* DIAP-1 (reviewed in [25]), as well as IAP-like 2 (ILP2) [26], as caspase inhibitors. IAPs might have functions other than caspase inhibition because they have been found in organisms such as yeast, which neither contain caspases nor undergo apoptosis [27].

IAPs contain one, two, or three BIR (baculovirus IAP repeat) domains, which represent the defining characteristic of the family. At present, there is no known function for BIR1; however, domains closely related to the second BIR domain (BIR2) of XIAP specifically target caspases 3 and 7, and regions closely related to the third BIR domain (BIR3) specifically target caspase 9 (reviewed in [24,28]). This led to the general assumption that the BIR domain itself was important for caspase inhibition. Surprisingly, the recently reported structures of BIR2 in complex with caspases 3 and 7 have revealed the BIR domain to have almost no direct role in the inhibitory mechanism. All the important inhibitory contacts are made by the flexible region preceding the BIR domain (reviewed in [28]).

Interestingly, in addition to these endogenous regulators, the cowpox virus CrmA protein and the baculovirus p35 protein are produced early in infection to suppress caspase-mediated host responses. Each of the inhibitors has a characteristic specificity profile against human caspases (reviewed in [29]). Though XIAP and CrmA would be expected to regulate mammalian caspases *in vivo*, p35 would never be present normally in mammals because it is expressed naturally by baculoviruses. Many homologues of CrmA (serpins) exist in humans, but none of them inhibit caspases efficiently. Interestingly, no endogenous homologue of baculovirus p35 has been detected, in humans or even in baculovirus hosts.

## Perspectives

The basic order and at least some of the essential functions and, importantly, endogenous regulators of the caspases are known. But this core pathway probably represents a minimal apoptotic programme, and certainly its simplicity is complicated by cell-specific additions that help to fine-tune individual cell fates. Part of the cell specificity of the apoptotic programme comes from the availability of specific caspase substrates [30], and the challenge here is to elucidate whether cleavage of the several hundred presumptive caspase substrates is part of the apoptotic programme, or a post-mortem event. Another goal for the future is to understand the cell-specific regulation of caspase and caspase inhibitor concentrations in individual cells. Having grounded the field in the basic biochemistry of the system, the big job now is to explore and understand these cell-specific controls.

## Summary

- *Caspases are essential for cytokine activation and apoptosis.*
- *Different caspase family members are involved in the distinct biologies of apoptosis and cytokine activation, and this is dictated in part by their extended substrate specificity.*
- *The distinctive property of all members is an exquisite specificity for aspartate in the P1 position of substrates.*
- *Apoptosis is driven by the zymogen-to-active-form transition of caspases 8 and 9 at distinct macromolecular assemblies, and this initiation phase is followed by direct activation of the zymogens of caspases 3 and 7.*
- *Caspase activity is under the control of endogenous inhibitors belonging to the IAP family.*


Work reported in this essay was enabled by grants from the National Institutes of Health. I apologize for not citing many pertinent papers owing to publisher's limitations.

## References


1. Chen, J.M., Rawlings, N.D., Stevens, R.A. & Barrett, A.J. (1998) Identification of the active site of legumain links it to caspases, clostripain and gingipains in a new clan of cysteine endopeptidases. *FEBS Lett.* **441**, 361–365
2. Zheng, T.S., Hunot, S., Kuida, K. & Flavell, R.A. (1999) Caspase knockouts: matters of life and death. *Cell Death Differ.* **6**, 1043–1053
3. Lippens, S., Kockx, M., Knaapen, M., Mortier, L. Polakowska, R., Verheyen, A., Garmyn, M., Zwijsen, A., Formstecher, P., Huylebroeck, D. et al. (2000) Epidermal differentiation does not involve the pro-apoptotic executioner caspases, but is associated with caspase-14 induction and processing. *Cell Death Differ.* **7**, 1218–1224
4. Hofmann, K., Bucher, P. & Tschopp, J. (1997) The CARD domain: a new apoptotic signalling motif. *Trends Biochem. Sci.* **22**, 155–156
5. Chinnaiyan, A.M., O'Rourke, K., Tewari, M. & Dixit, V.M. (1995) FADD, a novel death domain-containing protein, interacts with the death domain of Fas and initiates apoptosis. *Cell* **81**, 505–512
6. Wei, Y., Fox, T., Chambers, S.P., Sintchak, J., Coll, J.T., Golec, J.M., Swenson, L., Wilson, K.P. & Charifson, P.S.. (2000) The structures of caspases-1, -3, -7 and -8 reveal the basis for substrate and inhibitor selectivity. *Chem. Biol.* **7**, 423–432
7. Salvesen, G. (1999) Caspase 8: igniting the death machine. *Structure* **7**, 225–229
8. Renatus, M., Stennicke, H.R., Scott, F.L., Liddington, R.C. & Salvesen, G.S. (2001) A self-priming mechanism drives the activation of the cell death protease caspase 9. *Proc. Natl. Acad. Sci. U.S.A.* **98**, 14250–14255
9. Garcia-Calvo, M., Peterson, E.P., Rasper, D.M., Vaillancourt, J.P., Zamboni, R., Nicholson, D.W. & Thornberry, N.A. (1999) Purification and catalytic properties of human caspase family members. *Cell Death Differ.* **6**, 362–369
10. Stennicke, H.R. & Salvesen, G.S. (1997) Biochemical characteristics of caspases-3, -6, -7, and -8. *J. Biol. Chem.* **272**, 25719–25723
11. Stennicke, H.R., Renatus, M., Meldal, M. & Salvesen, G.S. (2000) Internally quenched fluorescent peptide substrates disclose the subsite preferences of human caspases 1, 3, 6, 7 and 8. *Biochem. J.* **350**, 563–568

12. Thornberry, N.A., Rano, T.A., Peterson, E.P., Rasper, D.M., Timkey, T., Garcia-Calvo, M., Houtzager, V.M., Nordstrom, P.A., Roy, S., Vaillancourt, J.P. et al. (1997) A combinatorial approach defines specificities of members of the caspase family and granzyme B. Functional relationships established for key mediators of apoptosis. *J. Biol. Chem.* **272**, 17907–17911
13. Cameron, P., Limjuco, G., Rodkey, J., Bennett, C. & Schmidt, J.A. (1985) Amino acid sequence analysis of human interleukin 1 (IL-1). Evidence for biochemically distinct forms of IL-1. *J. Exp. Med.* **162**, 790–801
14. Lazebnik, Y.A., Kaufmann, S.H., Desnoyers, S., Poirier, G.G. & Earnshaw, W.C. (1994) Cleavage of poly(ADP-ribose) polymerase by a proteinase with properties like ICE. *Nature (London)* **371**, 346–347
15. Liu, X., Zou, H., Slaughter, C. & Wang, X. (1997) DFF, a heterodimeric protein that functions downstream of caspase-3 to trigger DNA fragmentation during apoptosis. *Cell* **89**, 175–184
16. Casciola-Rosen, L.A., Miller, D.K., Anhalt, G.J. & Rosen, A. (1994) Specific cleavage of the 70-kDa protein component of the U1 small nuclear ribonucleoprotein is a characteristic biochemical feature of apoptotic cell death. *J. Biol. Chem.* **269**, 30757–30760
17. Brockstedt, E., Rickers, A., Kostka, S., Laubersheimer, A., Dorken, B., Wittmann-Liebold, B., Bommert, K. & Otto, A. (1998) Identification of apoptosis-associated proteins in a human Burkitt lymphoma cell line. Cleavage of heterogeneous nuclear ribonucleoprotein A1 by caspase 3. *J. Biol. Chem.* **273**, 28057–28064
18. Otto, A., Muller, E.C., Brockstedt, E., Schumann, M., Rickers, A., Bommert, K. & Wittmann-Liebold, B. (1998) High performance two dimensional gel electrophoresis and nanoelectrospray mass spectrometry as powerful tool to study apoptosis-associated processes in a Burkitt lymphoma cell line. *J. Protein Chem.* **17**, 564–565
19. Stennicke, H.R., Deveraux, Q.L., Humke, E.W., Reed, J.C., Dixit, V.M. & Salvesen, G.S. (1999) Caspase-9 can be activated without proteolytic processing. *J. Biol. Chem.* **274**, 8359–8362
20. Salvesen, G.S. and Dixit, V.M. (1999) Caspase activation: the induced-proximity model. *Proc. Natl. Acad. Sci. U.S.A.* **96**, 10964–10967
21. Chai, J., Wu, Q., Shiozaki, E., Srinivasula, S.M., Alnemri, E.S. & Shi, Y. (2001) Crystal structure of a procaspase-7 zymogen. Mechanisms of activation and substrate binding. *Cell* **107**, 399–407
22. Riedl, S.J., Fuentes-Prior, P., Renatus, M., Kairies, N., Krapp, R., Huber, R., Salvesen, G.S. & Bode, W. (2001) Structural basis for the activation of human procaspase-7. *Proc. Natl. Acad. Sci. U.S.A.* **98**, 14790–14795
23. Tschopp, J., Irmler, M. & Thome, M. (1998) Inhibition of fas death signals by FLIPs. *Curr. Opin. Immunol.* **10**, 552–558
24. Deveraux, Q.L. & Reed, J.C. (1999) IAP family proteins – suppressors of apoptosis. *Genes Dev.* **13**, 239–252
25. Verhagen, A.M., Coulson, E.J. & Vaux, D.L. (2001) Inhibitor of apoptosis proteins and their relatives: IAPs and other BIRPs. *Genome Biol.* **2**, reviews3009
26. Richter, B.W., Mir, S.S., Eiben, L.J., Lewis, J., Reffey, S.B., Frattini, A., Tian, L., Frank, S., Youle, R.J., Nelson, D.L. et al. (2001) Molecular cloning of ILP-2, a novel member of the inhibitor of apoptosis protein family. *Mol. Cell Biol.* **21**, 4292–4301
27. Uren, A.G., Coulson, E.J. & Vaux, D.L. (1998) Conservation of baculovirus inhibitor of apoptosis repeat proteins (BIRPs) in viruses, nematodes, vertebrates and yeasts. *Trends Biochem. Sci.* **23**, 159–162
28. Fesik, S.W. & Shi, Y. (2001) Structural biology: controlling the caspases. *Science* **294**, 1477–1478
29. Stennicke, H.R., Ryan, C.A. & Salvesen, G.S. (2002) Reprieval from execution: the molecular basis of caspase inhibition. *Trends Biochem. Sci.* **27**, 94–101
30. Nicholson, D.W. (1999) Caspase structure, proteolytic substrates, and function during apoptotic cell death. *Cell Death Differ.* **6**, 1028–1042
31. Zou, H., Li, Y., Liu, X. & Wang, X. (1999) An APAF-1.cytochrome c multimeric complex is a functional apoptosome that activates procaspase-9. *J. Biol. Chem.* **274**, 11549–11556

3

# Matrix metalloproteinases in cancer

Yoshifumi Itoh[1] and Hideaki Nagase

*Kennedy Institute of Rheumatology, Faculty of Medicine, Imperial College of Science, Technology and Medicine, 1 Aspenlea Road, Hammersmith, London W6 8LH, U.K.*

## Abstract

The extracellular matrix (ECM) holds cells together and maintains the three-dimensional structure of the body. It also plays critical roles in cell growth, differentiation, survival and motility. For a tumour cell to metastasize from the primary tumour to other organs, it must locally degrade ECM components that are the physical barriers for cell migration. The key enzymes responsible for ECM breakdown are matrix metalloproteinases (MMPs). To date, 23 MMP genes have been identified in humans and many are implicated in cancer. ECM degradation by MMPs not only enhances tumour invasion, but also affects tumour cell behaviour and leads to cancer progression. This review highlights recent developments with regard to the cellular and molecular mechanisms of MMPs that influence tumour cell growth, invasion and metastasis.

## Introduction

Multicellular organisms consist of cells and a complex network of extracellular macromolecules such as collagens, proteoglycans, fibronectin, lamins and many other glycoproteins. This network, referred to as the extracellular matrix (ECM), holds cells together in an organized assembly, guides cell migration and creates correct cellular environments. The ECM also acts as a reservoir of growth factors and provides signals to the cell through ECM receptors on the

[1]*To whom correspondence should be addressed (e-mail: y.itoh@ic.ac.uk).*

cell surface [1]. The ECM thus plays essential roles in many biological processes, e.g. embryonic development, morphogenesis, tissue resorption and repair, cell differentiation, migration, growth and apoptosis. Degradation of the ECM modifies not only the structure of tissue but also cellular function and behaviour. The activities of ECM-degrading proteinases must therefore be precisely regulated. Although many proteinases are implicated in ECM degradation, a group of metalloproteinases called matrix metalloproteinases (MMPs), or matrixins, is considered to play a major role. ECM turnover associated with uncontrolled matrixin activities is involved in diseases such as arthritis, atherosclerosis, fibrosis and cancer. The involvement of MMPs in cancer metastasis has attracted particular attention, since it raises the possibility of developing MMP inhibitors as a new generation of cancer treatment. In addition to the involvement in metastasis, recent studies indicate that MMPs are also involved in vascularization and initial tumour development. In this review we discuss the current understanding of the role of MMPs in cancer metastasis and tumour progression.

## What are the MMPs?

MMPs are structurally related zinc metalloproteinases (proteinases that contain a zinc atom at the catalytic site that is essential for hydrolysis of a peptide bond; see Chapter 1). They are secreted from the cell (soluble MMPs) or bound to the cell surface (membrane-type MMPs; TMT-MMPs) and degrade ECM and other proteins (Table 1) [2,3]. At present, 23 mammalian MMPs have been identified and they are classified according to their substrate specificity and structural similarity. All MMPs share common domain structures including a signal sequence, a propeptide, a catalytic domain, and a hemopexin-like (Hpx) domain (Figure 1). Propeptides contain a unique sequence signature called the 'cysteine switch' with a PRCGXPD motif, whose cysteine residue (underlined) interacts with the catalytic zinc in the catalytic domain as a fourth ligand, thereby keeping the precursor zymogen (the inactive precursor form of an enzyme) proMMP inactive. Catalytic domains have a zinc-binding motif HEXGHXXGXXH, in which the three histidine residues are ligands of the catalytic zinc atom. The two gelatinases (MMP-2 and MMP-9) have three additional repeats of a fibronectin type II-like domain inserted in the catalytic domain, which enables them to bind to collagen. The Hpx domain often plays an important role in protein–protein interactions and in determining enzyme specificities. For collagenases to recognize and cleave triple-helical collagens, the Hpx domain is an absolute requirement. Two matrilysins (MMP-7 and MMP-26) lack the Hpx domain, and MMP-23 has a unique cysteine array and Ig-like domain instead of the Hpx domain.

MT-MMPs are further sub-divided into a transmembrane type and a glycosylphosphatidylinositol (GPI)-anchored type. Transmembrane type MT-MMPs

**Table 1. Substrates of mammalian MMPs**

*Two identical genes are found in a head-to-head arrangement in chromosome 1. $\alpha_1$PI, $\alpha_1$-proteinase inhibitor; $\alpha_2$M, $\alpha_2$-macroglobulin; $\alpha_1$-ACT, $\alpha_1$-antichymotrypsin; TNF$\alpha$, tumour necrosis factor $\alpha$; IGFBP-3, insulin-like growth factor binding molecule 3; MCP-3, monocyte chemoattractant protein 3; SPARC, secreted protein acidic and rich in cysteine (osteonectin, BM40); COMP, cartilage oligomeric matrix protein; IL-1$\beta$, interleukin 1$\beta$; HSPG, heparan sulphate proteoglycan; DSPG, dermatan sulphate proteoglycan; C1q, complement protein 1q.

| Enzyme | MMP | Substrates |
|---|---|---|
| **Soluble types** | | |
| *Collagenases* | | |
| Interstitial collagenase (collagenase 1) | MMP-1 | Collagens I, II, III, VII, VIII, X, and XI, gelatin, C1q, entactin, tenascin, aggrecan, link protein, fibronectin, vitronectin, myelin basic protein, IGFBP-3, casein, $\alpha_1$PI, $\alpha_1$-ACT, IL-1$\beta$, $\alpha_2$M, proTNF$\alpha$ |
| Neutrophil collagenase (collagenase 2) | MMP-8 | Collagens I, II, and III, C1q, aggrecan, $\alpha_1$PI, substrate P, $\alpha_2$M |
| Collagenase 3 | MMP-13 | Collagens I, II, III, IV, IX, X, and XIV, gelatin, collagen telopeptides, C1q, fibronectin, SPARC, aggrecan, casein, $\alpha_2$M |
| *Gelatinases* | | |
| Gelatinase A | MMP-2 | Collagens I, II, III, IV, V, VII, and X, gelatin, fibronectin, laminin, aggrecan, link protein, elastin, vitronectin, tenascin, SPARC, docorin, myelin basic protein, $\alpha_1$PI, $\alpha_1$-ACT, IL-1$\beta$, IGFBP-3, substance P, $\alpha_2$M, proTNF$\alpha$, MCP-3 |
| Gelatinase B | MMP-9 | Collagens IV, V, XI, XIV, elastin, aggrecan, link protein, decorin, laminin, entactin, SPARC, myelin basic protein, $\alpha_1$PI, IL-1$\beta$, substance P, casein, $\alpha_2$M, proTNF$\alpha$ |

**Table 1 (contd.)**

| Enzyme | MMP | Substrates |
| --- | --- | --- |
| *Stromelysins* | | |
| Stromelysin 1 | MMP-3 | Collagens III, IV, V, IX, X, and XI, teropeptides (collagen I and II), gelatin, aggrecan, link protein, elastin, fibronectin, vitronectin, laminin, entactin, tenascin, SPARC, decorin, myelin basic protein, $\alpha_1$PI, $\alpha_1$-ACT, IL-1$\beta$, IGFBP-3, substance P, T kininogen, casein, proMMP-1, proMMP-3, proMMP-8, proMMP-9, $\alpha_2$M |
| Stromelysin 2 | MMP-10 | Collagen III, IV, and V, gelatin, fibronectin, elastin, aggrecan, link protein, casein, proMMP-1, proMMP-7, proMMP-8, proMMP-9, $\alpha_2$M |
| *Matrilysins* | | |
| Matrilysin 1 | MMP-7 | Collagen IV, gelatin, aggrecan, link protein, elastin, fibronectin, vitronectin, laminin, SPARC, entactin, decorin, myelin basic protein, tenascin, fibulin, casein, $\alpha_1$PI, proMMP-1, proMMP-2, proMMP-9, $\alpha_2$M, proTNF$\alpha$ |
| Matrilysin 2 | MMP-26 | Fibronectin, fibrinogen, vitronectin, gelatin, $\alpha_1$-PI, $\alpha_2$M |
| *Others* | | |
| Stromelysin 3 | MMP-11 | $\alpha_1$PI, $\alpha_2$M, (for mouse enzyme: collagen IV, gelatin, fibronectin, laminin, aggrecan) |
| Metalloelastase | MMP-12 | Elastin, collagen IV, gelatin, fibronectin, vitronectin, laminin, entactin, aggrecan, myelin basic protein, $\alpha_2$M, $\alpha_1$PI, proTNF$\alpha$ |

| No name | MMP-19 | Collagen type IV, laminin, nidogen, large tenascin-C isoform, fibronectin, gelatin, aggrecan, COMP |
|---|---|---|
| Enamelysin | MMP-20 | Amerogenin, COMP |
| CA-MMP | MMP-23* | Gelatin |
| No name | MMP-27 | Not known |
| Epilysin | MMP-28 | Casein |
| | | |
| **Membrane types** | | |
| *Transmembrane* | | |
| MT1-MMP | MMP-14 | ProMMP-2, proMMP-13, collagens I, II, and III, gelatin, fibronectin, vitronectin, laminins 1 and 5, entactin, aggrecan, fibrin, $\alpha_2$M, $\alpha_1$PI, decorin, proTNF$\alpha$, CD44H |
| MT2-MMP | MMP-15 | ProMMP-2, laminin, fibronectin, tenascin, entactin, aggrecan, perlecan, proTNF$\alpha$ |
| MT3-MMP | MMP-16 | ProMMP-2, collagen III, fibronectin |
| MT5-MMP | MMP-24 | ProMMP-2, HSPG, DSPG, gelatin |
| | | |
| *GPI-anchored* | | |
| MT4-MMP | MMP-17 | Fibrinogen, fibrin, proTNF$\alpha$ |
| MT6-MMP | MMP-25 | Collagen IV, gelatin, fibrinogen, X-linked fibrin, fibronectin |

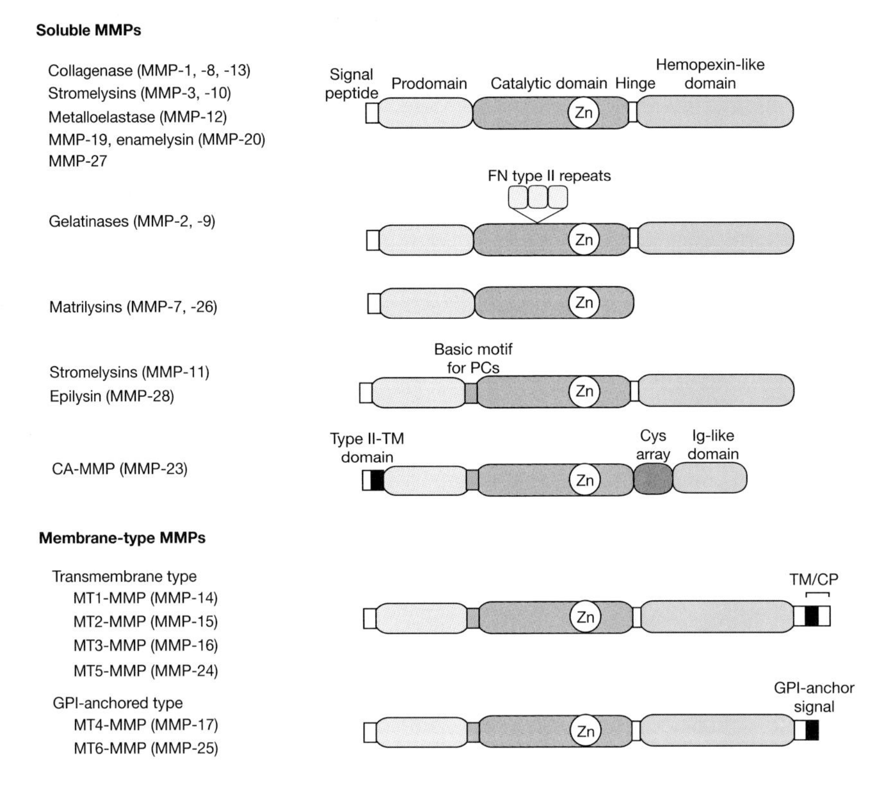


**Figure 1. Domain structures of MMPs**

Schematic representations of the domain structures of MMPs are shown. FN type II, fibronectin type II-like domain; PC, pro-hormone convertase; Cys array, cysteine array domain; TM/CP, transmembrane and cytoplasmic domains.

have a transmembrane domain and a short cytoplasmic tail at the C-terminus. GPI-anchored type MT-MMPs have a stretch of hydrophobic GPI-anchoring signal sequence at the C-terminus [4].

Most of the soluble MMPs are activated extracellularly by proteinases. Exceptions, however, are MMP-11, MMP-23 and MMP-28, which are thought to be activated intracellularly and secreted in an active form. These MMPs have the basic amino acid motif [RX(R/K)R] at the end of the propeptide, facilitating recognition by proprotein convertases, such as furin, in the Golgi apparatus (see Chapter 7). MMP-23 has a type II transmembrane domain in the N-terminal propeptide, so it becomes a soluble enzyme upon activation. MT-MMPs also harbour the RX(R/K)R motif at the end of propeptide, so that they are also likely to be processed by a proprotein convertase and appear on the cell surface as the active form.

Table 1 lists the protein substrates degraded by MMPs that have been characterized mostly *in vitro*. Some MMPs, such as MMP-11 and MT4-MMP (MMP-17), have only a weak proteolytic activity against ECM proteins. The activities of the more recently discovered MMP-27 and MMP-28 have not been well characterized. Recent studies have indicated that the functions of MMPs are not simply to destroy the ECM, but to expose cryptic functions of ECM molecules through proteolysis, e.g. specific cleavage of $\gamma 2$ chain of laminin 5 by MMP-2 or MT1-MMP stimulates epithelial cell migration (see below). MMPs can also process soluble non-ECM proteins that exhibit biological activities, e.g. inactivation of interleukin 1$\beta$, monocyte chemoattractant protein 3 and insulin-like growth factor binding proteins, and the activation of tumour necrosis factor $\alpha$.

The activities of MMPs are regulated by endogenous inhibitors, plasma inhibitor $\alpha_2$-macroglobulin and tissue inhibitors of metalloproteinases (TIMPs) [5]. $\alpha_2$-Macroglobulin inhibits most endopeptidases (proteolytic enzymes that hydrolyse the internal peptide bond of a polypeptide chain; see Chapter 1, Figure 3) and it is also an effective inhibitor of MMPs. TIMPs are the main inhibitors in tissues. Four TIMPs (TIMP-1, -2, -3 and -4) have been identified in mammals. They are 21–29 kDa proteins with six conserved disulphide bonds and are composed of N-terminal and the C-terminal domains (Figure 2). The MMP inhibitory site is located in the N-terminal domain, with residues 1–4 and 68–70 being the critical regions. $Cys^1$ and $Cys^{70}$ are linked by a disulphide bond, and form a wedge-like ridge that slots into the active site cleft of MMPs. The catalytic zinc is bound by the $\alpha$-amino group and the carbonyl oxygen of the N-terminal cysteine (Figure 2). TIMPs are specific inhibitors of MMPs and they do not inhibit other metalloproteinases. The exception is TIMP-3 which inhibits some members of the ADAM (a disintegrin and metalloproteinase) family, e.g. ADAM-10, -12, -17, and ADAMTS-4 and -5 (ADAM with thrombospondin motifs) (see Chapter 11).

Many synthetic MMP inhibitors have been generated in the hope that they can be used for treatments of MMP-related diseases, including cancer. They bind to the active site of the enzyme like a substrate and inhibit it by chelating the catalytic zinc atom with a moiety such as a hydroxamic acid, a thiol, a carboxyl, or a phosphorous group.

## MMPs and tumour cell invasion

Metastasis is the spread of cancer cells from the primary tumour to distant sites in the body. It is the leading cause of death in cancer patients. For a tumour cell to metastasize it must accomplish the following four events: (i) detachment from the primary tumour and subsequent invasion into the connective tissue stroma; (ii) entrance to the blood vessel or lymphatic system (intravasation) to traverse to distant sites in the body; (iii) exudation from the circulation (extavasation); and (iv) formation of metastatic colonies (Figure 3). For migratory cells to achieve these tasks, ECM components are major physical barriers. Liotta et al. [6] have proposed a three-step theory to explain how calls can overcome this barrier (Figure 4). First, the tumour cells attach to the matrix macromolecules of the basement membrane or stroma via specific cell-surface receptors. This is followed by local degradation of the ECM. At this point, the anchored tumour cell needs

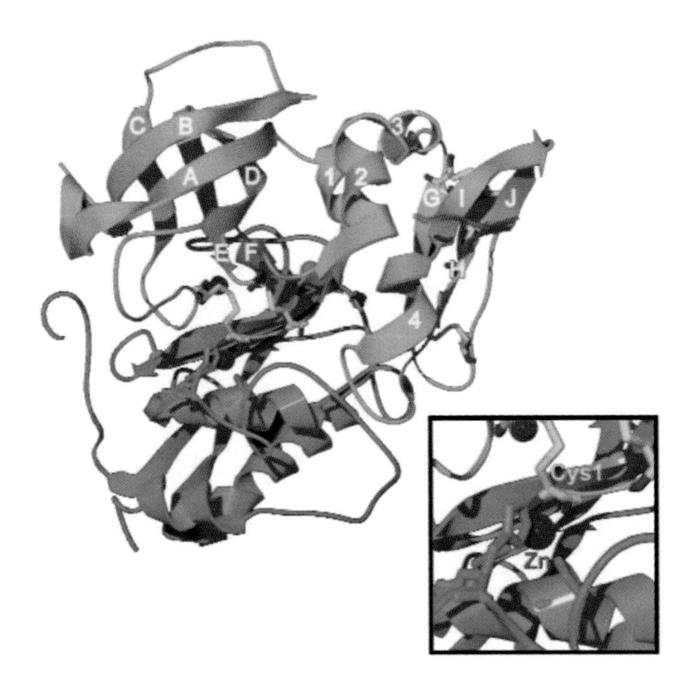


**Figure 2. MMP-3–TIMP-1 complex**
Ribbon representation of the structure of MMP-3 catalytic domain–TIMP-1 complex (Protein Data Bank entry 1UEA) [23] where MMP-3 is shown in blue and TIMP-1 in orange. The catalytic domain of MMP-3 consists of three α-helices and five β-strands (sky blue), one catalytic and one structural zinc ions (pink) and three calcium ions (orange). TIMP-1 consists of N-terminal (contains α-helix 1, β-sheets A, B, C, and D, and α-helixes 2 and 3) and C-terminal (contains β-sheets G and H, α-helix 4, and β-sheets I and J) domains. The N-terminal inhibitory domain forms a β-barrel structure similar to that of the oligosaccharide/oligonucleotide-binding (OB)-fold family proteins. Six conserved disulphide bonds are shown in yellow. An enlarged image of the complex around the catalytic zinc atom of the MMP-3 is shown in the inset. The catalytic zinc atom (pink) is held by three histidine residues [$His^{201}$, $His^{205}$ and $His^{211}$ (sky blue)] in MMP-3. The α-amino group (blue) and the carbonyl oxygen (red) group of the N-terminal $Cys^1$ in TIMP-1 chelate the zinc atom of the active site (pink). The figure was prepared with the SWISS PDB viewer [24] and rendered by POV-Ray software.

to utilize proteolytic enzymes, which may arise from the tumour cell itself or come from the surrounding stroma. Finally, the cell migrates towards the degraded ECM, which dictates the directionality of cell locomotion. Continued invasion of tumour cells into the tissue is achieved by cyclic repetition of these three steps. It is now widely accepted that MMPs play the major role in breaking down the ECM and creating a path for cancer cells. There have been numerous reports showing a correlation between increased expression of various MMPs and tumour progression in a range of cancer cells as well as in surrounding stromal cells [7]. These include MMP-1 (collagenase 1), MMP-2 (gelatinase A), MMP-3 (stromelysin 1), MMP-7 (matrilysin 1), MMP-9 (gelatinase B), MMP-10 (stromelysin 2), MMP-11 (stromelysin 3), MMP-13 (collagenase 3), and MT1-MMP (MMP-14). The involvement of MMPs in cell invasion is also supported by

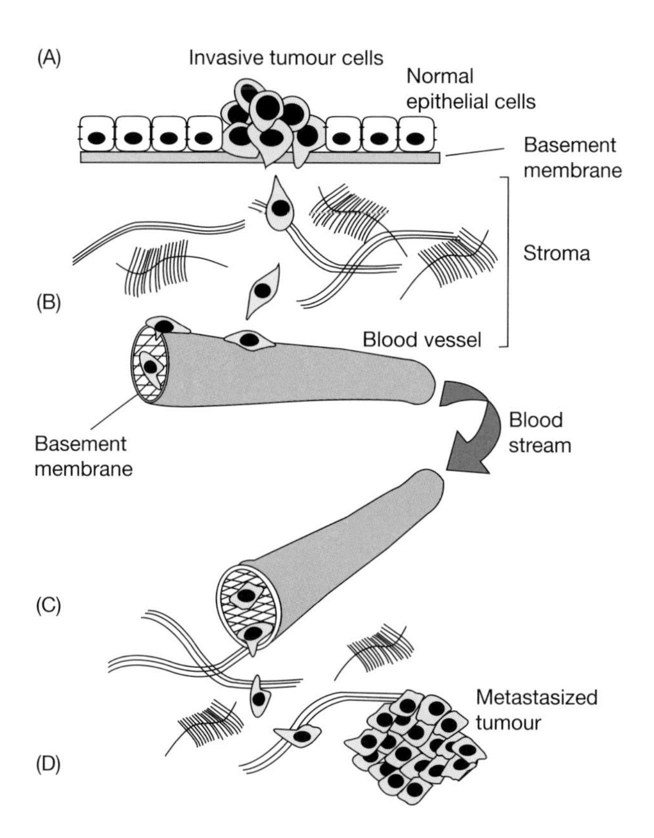


**Figure 3. Processes involved in cancer cell metastasis**
The metastatic process of the cancer cell derived from epithelial cells. In step (A), transformed cells grow and start to invade the stromal tissue. This requires a decrease in cell–cell contact and breakdown of the basement membrane to enable the cell to invade the stromal tissue. The cells then reach the vasculature, break its basement membrane, and enter the blood stream (step B). When the cells reach the destination tissue, they attach to the endothelial cells, degrade the basement membrane, and invade the stromal tissue (step C). The successfully metastasized cells re-grow and form metastatic colonies, with associated angiogenesis.

the series of inhibitor studies using TIMPs or synthetic MMP inhibitors *in vivo* as well as *in vitro*. However, it has not been clear which MMP or MMPs are functionally involved in the invasive process *in vivo*. It is likely that multiple MMPs are involved and that these may vary in different types of cancer.

## Importance of pericellular proteolysis in tumour cell invasion

When cells invade a tissue, it is not necessary to degrade a large area of ECM, but a more focal pericellular area (the area immediately surrounding cell) in the direction of the migration. Thus, proteinases that are bound to the plasma membrane are likely to be more suitable for this purpose than soluble enzymes. Hotary et al. [8] reported that the overexpression of soluble MMPs, including MMP-1, -3, -7, -9, -11, and –13, did not modify the invasion ability

**Step 1. Attachment**

Attachment through specific ECM-receptors

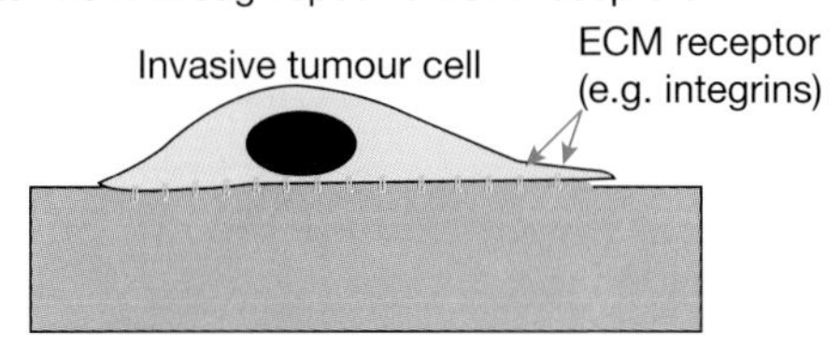


**Step 2. ECM degradation**

Focal degradation of ECM by MMPs

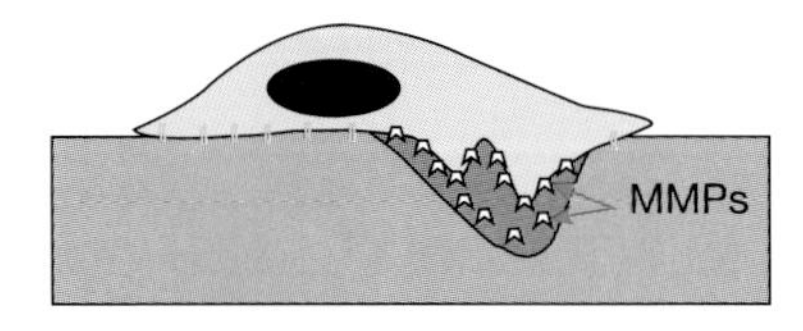


**Step 3. Locomotion**

Cell migration into the degradation region of ECM

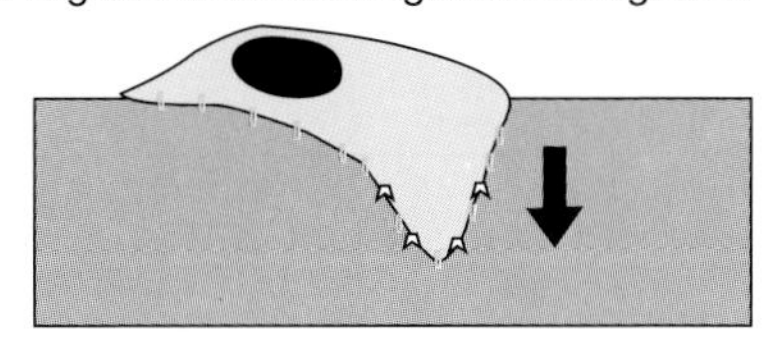

**Figure 4. Molecular events that take place during cell invasion: three-step theory**
The cell attaches to the ECM via specific ECM receptors on the cell surface (Step 1). The cell then starts to express proteinases (MMPs) that degrade ECM at the attachment site (Step 2). It then migrates to the area where the ECM was degraded (Step 3). Repetition of these steps results in cellular invasion.

of Madin–Darby canine kidney epithelial cells into a collagen matrix, whereas overexpression of membrane-bound MT1-MMP did. Yu and Stamenkovic [9] also showed that a membrane-anchored chimaera mutant of MMP-9 and the transmembrane and cytoplasmic domains of CD44 (a cell surface adhesion molecule that binds to ECM components including hyaluronan, collagens and fibronectin) enhanced cellular invasion into the matrix compared with the wild-type soluble form. Despite the fact that the expression of many soluble MMPs is upregulated at the invasive sites of tumours, these studies emphasize that pericellular proteolysis is closely associated with the invasive phenotype and malignancy of cancer cells. Indeed, MT1-MMP has been shown to be one of the key enzymes in this process. This enzyme is found on the cell surface in its active form and degrades local ECM components [2,3]. MT1-MMP also activates the zymogens of MMP-2 and MMP-13, which degrade different ECM components at or near the cell surface [3,10]. The activation of MMP-2 is considered to be especially important, as MMP-2, but not MT1-MMP, can degrade type IV collagen, a major component of the basement membrane. In addition, both MT1-MMP and MMP-2 can cleave the $\gamma 2$ chain of laminin 5, another major component of the basement membrane. The resulting fragments promote the migration of epithelial cells with a low invasive potential [11,12]. More recently, it has been shown that MT1-MMP sheds CD44 from the cell surface, an activity that is essential for migration of the MIA PaCa-2 pancreatic cancer cell line [13]. CD44 is a major hyaluronan receptor and is expressed in many malignant tumour cells. It is therefore proposed that these invasive tumour cells use the CD44/MT1-MMP system for migration *in vivo*.

### Activation mechanism of proMMP-2

The activation of proMMP-2 by MT1-MMP is thought to be an integral part of tumour cell invasion. The mechanism is not a simple bimolecular interaction of proMMP-2 and MT1-MMP; several protein–protein interactions are involved, with the endogenous inhibitor TIMP-2 functioning as a bridging molecule [2,3]. The initial step is the binding of TIMP-2 to active MT1-MMP on the cell surface. This interaction is between the active site of the enzyme and the N-terminal inhibitory domain of TIMP-2. The MT1-MMP–TIMP-2 complex then acts as a receptor to recruit proMMP-2 produced by surrounding stromal cells. This binding occurs through the C-terminal Hpx domain of proMMP-2 and the C-terminal domain of TIMP-2. Since the MT1-MMP component of the MT1-MMP–TIMP-2–proMMP-2 complex is inhibited by TIMP-2, it is postulated that a second molecule of MT1-MMP is required for the activation of proMMP-2. Many of the recent studies support this notion [3], and more recently, it has been demonstrated that MT1-MMP forms a complex through the Hpx domains to keep the 'receptor MT1-MMP' and the 'catalytic MT1-MMP' close together [14] (Figure 5). The homophilic complexes of MT1-MMP (intermolecular interaction of identical molecules) are formed exclusively at the lamellipodia (a large plasma membrane protrusion that extends from the leading edge of the cell), indicating that MMP-2 activation and ECM degradation take place at the migration front of invading cells. Earlier histological studies showed that an elevated MMP-2 mRNA level is primarily found in cancer stromal cells, but that the MMP-2 protein is often associated with the malignant epithelium. In retrospect, these phenomena can be explained by the active recruitment of proMMP-2 by cancer cells through the molecular assembly of proMMP-2, MT1-MMP and TIMP-2.

## MMPs and tumour neovascularization

In order for tumours to grow to the size of more than a few millimetres in diameter, neovascularization or angiogenesis (the formation of new blood vessels from existing ones) must take place to supply oxygen and nutrients. Vascularization of the tumour also facilitates the invasion of the blood stream by tumour cells. To achieve neovascularization, it is generally thought that tumour

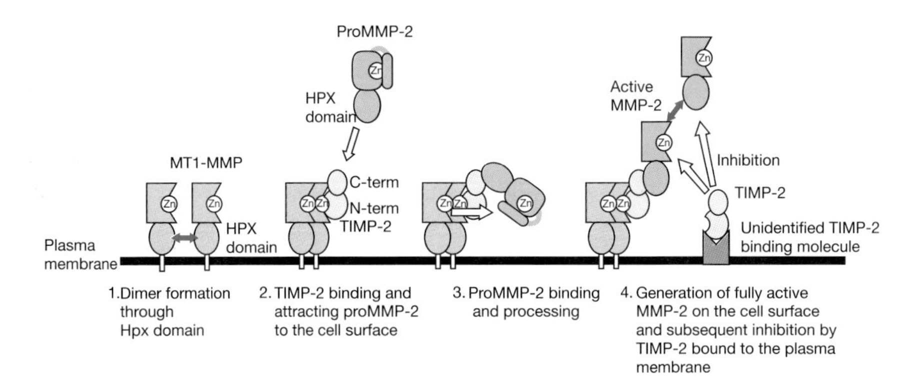


**Figure 5. Proposed activation mechanism of proMMP-2 by MT1-MMP on the cell surface**

Step 1: MT1-MMP molecules form a dimer complex through their Hpx domains. Step 2: TIMP-2 binds to one of the MT1-MMP molecules, which in turn attracts proMMP-2 to the cell surface through the interaction of the C-terminal domain of TIMP-2 and the Hpx domain of proMMP-2. Step 3: proMMP-2 bound to TIMP-2 is proteolytically activated by the other MT1-MMP. The cleavage occurs at the $Asn^{37}$–$Leu^{38}$ bond and generates an intermediate form of MMP-2, which becomes fully active by the complete autolytic removal of the rest of the propeptide. Step 4: active MMP-2 generated can stay on the cell surface to degrade cell periphery ECM or released from the cell surface to degrade a wider area of ECM. Certain populations of MMP-2 were shown to be inhibited by TIMP-2 bound on the cell surface through an unidentified molecule which is thought to be one of the effective regulatory mechanisms of MMP-2 after activation.

cells produce angiogenic factors such as vascular endothelial cell growth factor (VEGF) or fibroblast growth factor (FGF) to stimulate local neovascularization. However, Bergers et al. [15] have found that VEGF and FGF-1 (acidic FGF) are present in both the normal tissue of control mice and in the tumour tissue of a transgenic mouse model; however, neurovascularization is only observed in the tumour tissue. Furthermore, in this tumour model there were no differences in the expression of two VEGF receptors (Flk-1 and Flt-1) before and after angiogenesis took place, although VEGF signalling was still essential for angiogenesis. Therefore, what regulates angiogenesis in tumour tissue? What are the differences before and after angiogenesis? These researchers found that MMP-9 expression was upregulated in the tumour tissue and that it released VEGF that was bound to the ECM. The increased availability of VEGF activated quiescent endothelial cells to the angiogenic state.

VEGF increases vascular permeability, causing the leakage of blood proteins from the vascular bed. This is accompanied by activation of the blood clotting cascade and the formation of cross-linked fibrins. The deposited fibrin network then serves as a scaffold for angiogenic endothelial cells, but at the same time the highly cross-linked fibrin structure is also a major barrier for endothelial cell migration. Therefore, angiogenic endothelial cells need to utilize fibrinolytic proteinases for sprouting and capillary formation. Plasminogen activator and plasmin, well characterized fibrolytic proteases, were the prime candidates for this process, but Hiraoka et al. [16] found that inhibition of these serine proteinases did not influence the ability of endothelial cells to invade into cross-linked fibrin and to form capillaries. This process was inhibited by the MMP inhibitor TIMP-2, but not by TIMP-1. TIMP-2 inhibits MT1-MMP, but TIMP-1 does not. Based on a series of additional experiments, it was concluded that membrane-bound MT1-MMP is the key fibrinolytic enzyme in this system and promotes angiogenesis [16].

## MMPs and tumour development

The involvement of MMPs in tumour invasion and angiogenesis has been demonstrated as we have just discussed, but a number of observations have suggested that MMPs also play a role in tumorigenesis at a much earlier stage. For example, attenuating the TIMP-1 production in mouse fibroblasts by antisense RNA made them tumorigenic and metastatic in nude mice [17]. Overexpression of TIMP-1 or TIMP-2 in melanoma cells markedly reduced not only the invasive ability of those cells, but also primary tumour growth *in vivo* [18,19]. The role of MMP-7 in the development of intestinal tumours was investigated using a mouse strain which spontaneously develops intestinal tumours. This mouse model closely mimics the hereditary human colon cancer syndrome, familial adenomatous polyposis. MMP-7 production is elevated in the human intestinal tumour. As expected, MMP-7 was also expressed in epithelial-derived tumour cells in these mice, but not in normal intestinal mucosa. When these mice were crossed with MMP-7-deficient mice, the generated MMP-7-null mice developed 58% less tumours and the size of the tumours was 20% smaller than those of the wild-type [20]. This is further supported by inhibition of tumorigenesis by a synthetic MMP inhibitor. The mechanism underlying this phenomenon is not clear, but the proteolytic activity of MMP-7 is likely to be required for tumour growth. Sternlicht et al. [21] also reported that the overexpression of active MMP-3 in the mouse mammary gland results in aggressive malignant mammary tumours that exhibit genomic alterations, suggesting that the changes of microenvironment by the overexpression of MMPs can lead to DNA damage, which results in tumour generation. It is likely that soluble MMPs are suitable for such activities, as they can modify a broader area of the ECM. It is possible that other soluble MMPs that are overexpressed at or near the tumour may play such a role.

## Conclusions

Cellular and molecular biological studies have indicated that MMPs play a crucial role in cancer cell invasion and migration, in neovascularization and in tumorigenesis. Therefore, MMPs are certainly a suitable molecular target for cancer treatment. In the past decade, a large number of MMP inhibitors have been synthesized and have been shown to be effective against cancer in animal models. Unfortunately, when several of these inhibitors were tested clinically on patients with advanced cancer, they were, disappointingly, not effective. Therefore, are MMPs still suitable target molecules for therapeutic intervention? Is the investigation of MMPs in cancer still valid? The answer to these questions are certainly "Yes". Zucker et al. [22] commented that the failure in clinical efficacy may be due to the trials being conducted with advanced cancer patients, and it is emphasized that MMPs are more likely to play a critical role in the earlier stages of disease progression. The involvement of individual MMPs in each stage of cancer progression should now be analysed critically. It is important to investigate at what stage of the disease MMP inhibitors can work effectively. In addition, MMP inhibitors often have broad specificity and inhibit other types of metalloproteinases that are also biologically important such as ADAM enzymes (see Chapter 11). Therefore, there is a need to develop more specific inhibitors of a particular MMP to reduce possible side effects (e.g. joint pain, stiffness, oedema and reduced mobility). An understanding of the detailed interactions between MMPs and other related molecules may help us to develop new ways of blocking the function of a particular MMP with greater specificity. The engineering of mutant TIMPs specific to each MMP may also be considered because TIMPs have poor inhibitory activity against ADAM enzymes. TIMPs, on the other hand, have cell growth regulatory activities distinct from their MMP inhibitory function, and therefore, the inhibition of MMP by selective TIMP variants may have unexpected consequences. Further biochemical, molecular and cell biological investigations of MMPs and TIMPs are essential for our understanding of the molecular events in cancer progression, which will generate innovative ideas to fight this deadly disease in the future.

## Summary

- *MMPs are the major group of proteinases that degrade ECM macromolecules.*
- *Cancer cells utilize MMPs to invade into tissue.*
- *MT1-MMP, which participates in pericellular proteolysis of ECM, is one of the key enzymes involved in the invasive process of cancer cells.*
- *MMP-9 expressed in tumour tissues triggers an angiogenic switch by releasing VEGF from the ECM.*
- *Endothelial cells utilize MT1-MMP as a fibrinolytic proteinase to promote capillary formation.*
- *MMP-7 is involved in intestinal tumour generation.*
- *Aberrant ECM degradation by MMP may cause cellular transformation and cancer.*

## References


1. Werb, Z. (1997) ECM and cell surface proteolysis: regulating cellular ecology. *Cell* **91**, 439–442
2. Nagase, H. & Woessner, Jr, J.F. (1999) Matrix metalloproteinases. *J. Biol. Chem.* **274**, 21491–21494
3. Seiki, M. (1999) Membrane-type matrix metalloproteinases. *APMIS* **107**, 137–143
4. Itoh, Y., Kajita, M., Kinoh, H., Mori, H., Okada, A. & Seiki, M. (1999) Membrane type 4 matrix metalloproteinase (MT4-MMP, MMP-17) is a glycosylphosphatidylinositol-anchored proteinase. *J. Biol. Chem.* **274**, 34260–34266
5. Brew, K., Dinakarpandian, D. & Nagase, H. (2000) Tissue inhibitors of metalloproteinases: evolution, structure and function. *Biochim. Biophys. Acta* **1477**, 267–283
6. Liotta, L.A., Thorgeirsson, U.P. & Garbisa, S. (1982) Role of collagenases in tumor cell invasion. *Cancer Metastasis Rev.* **1**, 277–288
7. Fingleton, B. & Matrisian, L.M. (2001) Matrix metalloproteinases in cancer. In *Matrix Metalloproteinase Inhibitors in Cancer Therapy* (Clendeninn, N.J. & Appelt, K., eds), pp. 85–112, Humana Press, Totowa, NJ
8. Hotary, K., Allen, E., Punturieri, A., Yana, I. & Weiss, S.J. (2000) Regulation of cell invasion and morphogenesis in a three-dimensional type I collagen matrix by membrane-type matrix metalloproteinases 1, 2, and 3. *J. Cell Biol.* **149**, 1309–1323
9. Yu, Q. & Stamenkovic, I. (2000) Cell surface-localized matrix metalloproteinase-9 proteolytically activates TGF-beta and promotes tumor invasion and angiogenesis. *Genes Dev.* **14**, 163–176
10. Sato, H., Takino, T., Okada, Y., Cao, J., Shinagawa, A., Yamamoto, E. & Seiki, M. (1994) A matrix metalloproteinase expressed on the surface of invasive tumour cells. *Nature (London)* **370**, 61–65
11. Giannelli, G., Falk-Marzillier, J., Schiraldi, O., Stetler-Stevenson, W.G. & Quaranta, V. (1997) Induction of cell migration by matrix metalloprotease-2 cleavage of laminin-5. *Science* **277**, 225–228
12. Koshikawa, N., Giannelli, G., Cirulli, V., Miyazaki, K. & Quaranta, V. (2000) Role of cell surface metalloprotease MT1-MMP in epithelial cell migration over laminin-5. *J. Cell Biol.* **148**, 615–624
13. Kajita, M., Itoh, Y., Chiba, T., Mori, H., Okada, A., Kinoh, H. & Seiki, M. (2001) Membrane-type 1 matrix metalloproteinase cleaves CD44 and promotes cell migration. *J. Cell Biol.* **153**, 893–904
14. Itoh, Y., Takamura, A., Ito, N., Maru, Y., Sato, H., Suenaga, N., Aoki, T. & Seiki, M. (2001) Homophilic complex formation of MT1-MMP facilitates proMMP-2 activation on the cell surface and promotes tumor cell invasion. *EMBO J.* **20**, 4782–4793
15. Bergers, G., Brekken, R., McMahon, G., Vu, T.H., Itoh, T., Tamaki, K., Tanzawa, K., Thorpe, P., Itohara, S., Werb, Z. & Hanahan, D. (2000) Matrix metalloproteinase-9 triggers the angiogenic switch during carcinogenesis. *Nat. Cell Biol.* **2**, 737–744
16. Hiraoka, N., Allen, E., Apel, I.J., Gyetko, M.R. & Weiss, S.J. (1998) Matrix metalloproteinases regulate neovascularization by acting as pericellular fibrinolysins. *Cell* **95**, 365–377
17. Khokha, R., Waterhouse, P., Yagel, S., Lala, P.K., Overall, C.M., Norton, G. & Denhardt, D.T. (1989) Antisense RNA-induced reduction in murine TIMP levels confers oncogenicity on Swiss 3T3 cells. *Science* **243**, 947–950
18. Montgomery, A.M., Mueller, B.M., Reisfeld, R.A., Taylor, S.M. & DeClerck, Y.A. (1994) Effect of tissue inhibitor of the matrix metalloproteinases-2 expression on the growth and spontaneous metastasis of a human melanoma cell line. *Cancer Res.* **54**, 5467–5473
19. Koop, S., Khokha, R., Schmidt, E.E., MacDonald, I.C., Morris, V.L., Chambers, A.F. & Groom, A.C. (1994) Overexpression of metalloproteinase inhibitor in B16F10 cells does not affect extravasation but reduces tumor growth. *Cancer Res.* **54**, 4791–4797
20. Wilson, C.L., Heppner, K.J., Labosky, P.A., Hogan, B.L. & Matrisian, L.M. (1997) Intestinal tumorigenesis is suppressed in mice lacking the metalloproteinase matrilysin. *Proc. Natl. Acad. Sci. U.S.A.* **94**, 1402–1407
21. Sternlicht, M.D., Lochter, A., Sympson, C.J., Huey, B., Rougier, J.P., Gray, J.W., Pinkel, D., Bissell, M.J. & Werb, Z. (1999) The stromal proteinase MMP3/stromelysin-1 promotes mammary carcinogenesis. *Cell* **98**, 137–146

22. Zucker, S., Cao, J. & Chen, W.T. (2000) Critical appraisal of the use of matrix metalloproteinase inhibitors in cancer treatment. *Oncogene* **19**, 6642–6650
23. Gomis-Rüth, F.X., Maskos, K., Betz, M., Bergner, A., Huber, R., Suzuki, K., Yoshida, N., Nagase, H., Brew, K., Bourenkov, G.P. et al. (1997) Mechanism of inhibition of the human matrix metalloproteinase stromelysin-1 by TIMP-1. *Nature (London)* **389**, 77–81
24. Guex, N. & Peitsch, M.C. (1997) SWISS-MODEL and the Swiss-PdbViewer: an environment for comparative protein modeling. *Electrophoresis* **18**, 2714–2723

# 4

# Proteolytic processing of the amyloid-β protein precursor of Alzheimer's disease

Janelle Nunan and David H. Small[1]

*Department of Pathology, University of Melbourne, Parkville, Victoria 3010, Australia*

## Abstract

The proteolytic processing of the amyloid-β protein precursor plays a key role in the development of Alzheimer's disease. Cleavage of the amyloid-β protein precursor may occur via two pathways, both of which involve the action of proteases called secretases. One pathway, involving β- and γ-secretase, liberates amyloid-β protein, a protein associated with the neurodegeneration seen in Alzheimer's disease. The alternative pathway, involving α-secretase, precludes amyloid-β protein formation. In this review, we describe the progress that has been made in identifying the secretases and their potential as therapeutic targets in the treatment or prevention of Alzheimer's disease.

## Introduction

Alzheimer's disease (AD) is the most prevalent form of dementia in the elderly, causing progressive memory loss, confusion and behavioural changes. The presence of extracellular amyloid deposits in the brain is a key pathological feature of AD. Amyloid occurs in the neuropil in the form of amyloid plaques and around cerebral blood vessels in the form of cerebral amyloid angiopathy [1]. The major protein component of amyloid in the AD

[1]*To whom correspondence should be addressed (e-mail: d.small@pathology.unimelb.edu.au).*

brain is the amyloid-β protein (Aβ) [2], a 4 kDa polypeptide that is derived from a much larger precursor known as the amyloid-β protein precursor (APP) [3] (Figure 1). According to the amyloid hypothesis, the accumulation of Aβ in the brain is an important step in the pathogenesis of AD.

In this review, we describe our current understanding of the proteases involved in the proteolytic processing of APP. Inhibitors or activators of these proteases are emerging as key targets for the treatment of AD. The identification of an effective therapeutic strategy would be of enormous benefit as AD is one of the most common diseases associated with aging.

## APP: structure and processing

The APP gene is found on chromosome 21 and encodes a type 1 transmembrane glycoprotein [3]. The gene contains 18 exons, and parts of exons 16 and 17 encode the Aβ region, which lies partially within the ectodomain and partially within the transmembrane domain of APP. Alternative splicing of the APP mRNA can yield a number of isoforms. All forms of APP consist of a large N-terminal ectodomain that contains a signal peptide, a cysteine-rich domain and an acidic region, a short transmembrane domain, and a C-terminal cytoplasmic tail (Figure 1). Some isoforms also contain a domain that shows identity with Kunitz-type protease inhibitors and

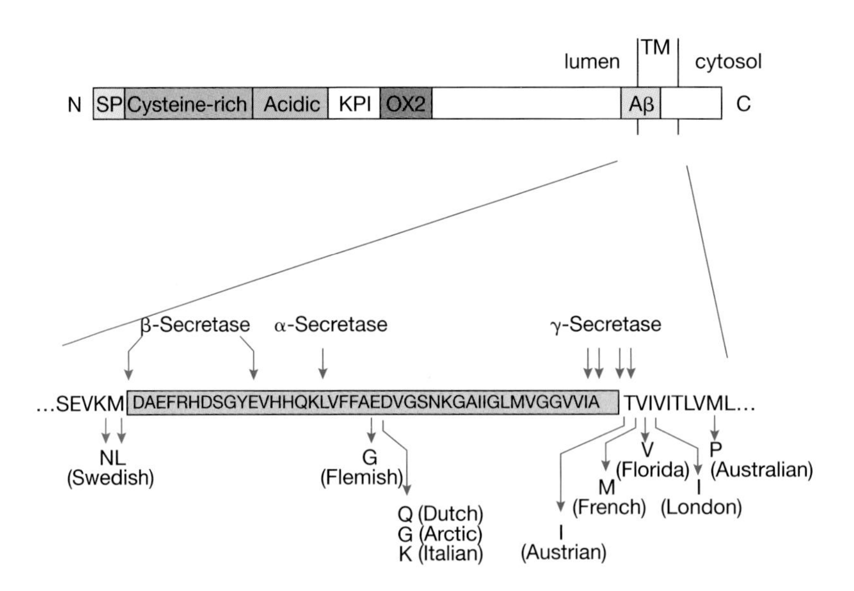


**Figure 1. Functional domains of APP and the amino-acid sequence of the Aβ region**
APP has a large ectodomain, with a signal peptide (SP), a cysteine-rich region, an acidic domain, a region with high identity with the Kunitz-type protease inhibitors (KPI) and a domain sharing identity with the OX-2 protein (OX2). The Aβ sequence lies partially within the ectodomain and partially within the transmembrane domain (TM). The cleavage sites for α-, β- and γ-secretases are shown, as are the mutations in APP that are known to cause FAD.

a region that shares identity with the OX-2 protein. In neurons, the major isoform contains 695 amino acid residues (APP695), and lacks the Kunitz-type protease inhibitor and OX-2 domains [4].

There are at least two pathways by which APP is proteolytically processed (reviewed in [5]). This processing results from the action of proteases that are termed 'secretases' (Figure 2). A major pathway involves cleavage by α-secretase, which cuts the Aβ sequence on the C-terminal side of $Lys^{16}$. This cleavage destroys the intact Aβ sequence. Alternatively, APP can be processed by β-secretase at the N-terminus of the Aβ sequence, producing a 99-amino-acid-residue C-terminal fragment, known as C99. Further processing of C99 by γ-secretase within the transmembrane region releases Aβ, which is approx. 40 amino-acid residues in length. Two major forms of Aβ are produced. Approx. 90% of the Aβ secreted contains 40 amino-acid residues (Aβ40), and a minor form, comprising approx. 10% of the total Aβ, contains 42 (Aβ42) or 43 (Aβ43) amino acids and is referred to as 'long Aβ'.

It is now generally believed that the accumulation of long Aβ in the brain of AD patients is intimately associated with neurodegeneration. Cell culture studies show that Aβ is toxic to neurons in culture and that the aggregation of the polypeptide into higher molecular mass oligomeric species is necessary for

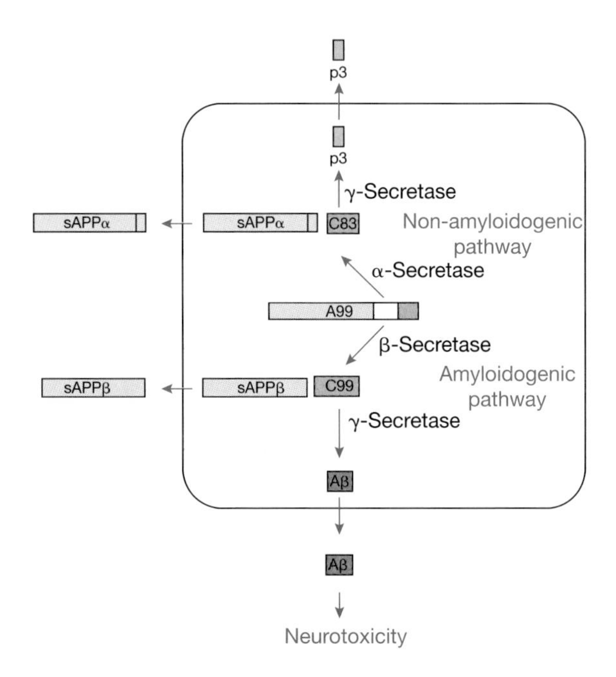


**Figure 2. Non-amyloidogenic and amyloidogenic pathways of APP processing**
Cleavage of APP by α-secretase produces sAPPα, which is secreted, and a C-terminal fragment C83. C83 is further processed by γ-secretase to release p3. The amyloidogenic pathway involves cleavage of APP by β-secretase to produce sAPPβ and C99, which is subsequently cleaved by γ-secretase to release Aβ. Aβ is secreted, and can form aggregates which have neurotoxic properties.

this effect [6]. As long Aβ aggregates more readily than Aβ40, these aggregates may contribute significantly to Aβ toxicity and amyloid plaque formation, by seeding the aggregation of the more abundant Aβ40.

## APP mutations

The strongest evidence that the production of Aβ is linked to the pathogenesis of AD comes from the identification of familial AD (FAD) mutations within the APP gene. FAD is inherited in an autosomal dominant manner, with the onset of clinical features usually occurring before the age of 65 years. In general, the phenotype and neuropathology of FAD is indistinguishable from late-onset 'sporadic' cases of the disease. This strongly suggests that the aetiologies of FAD and late-onset AD are similar.

A number of mutations in the APP gene have been identified which can cause early-onset FAD [4]. These mutations are clustered around the region that encodes Aβ (Figure 1). In general, all of these mutations have been found to increase the production of long Aβ (or an N-terminally truncated form of long Aβ). A double mutation located at codons 670 and 671 on the N-terminal side of the β-secretase cleavage site was first identified in a Swedish family. Another mutation at codon 692 near the α-secretase site was identified in a Flemish family. Several FAD mutations (London, Florida, French, Australian and Austrian) are clustered around the γ-secretase site within the transmembrane domain. These mutations change the site of γ-secretase cleavage to increase the proportion of long Aβ that terminates at amino acid residue 42 [7].

A reasonable strategy for AD therapy would be to inhibit Aβ toxicity, but the mechanism by which Aβ causes neurotoxicity is unclear [8]. Another strategy is to increase clearance of Aβ from the brain, and recent studies [9] suggest that a vaccination approach may be effective. However, at the time of writing, the safety of this approach and its efficacy are unknown. Considerable effort is now being expended on the design of compounds which inhibit Aβ production by acting on the β- and γ-secretases. In this review, we summarize our current knowledge about the α-, β- and γ-secretases and the compounds that inhibit or activate them.

## α-Secretase

α-Secretase provides a major route for APP processing, by cleaving on the C-terminal side of $Lys^{16}$ of the Aβ sequence (Figure 2). This cleavage produces a soluble N-terminal fragment (sAPPα) and a membrane-bound C-terminal fragment (C83). C83 undergoes further processing by γ-secretase, releasing a fragment known as p3.

α-Secretase has both a basal (constitutive) activity and a regulated activity, which is stimulated by protein kinase C (PKC) [4]. The activation of PKC via phorbol esters or via $M_1$ or $M_3$ muscarinic receptors increases α-secretase-mediated cleavage. PKC inhibitors reduce the activity of α-secretase to a residual level (the basal activity) [10]. The stimulation of the α-secretase pathway does not always cause a corresponding decrease in Aβ production, indicating that APP processing may vary between cell types.

Tumour necrosis factor-α (TNF-α)-converting enzyme (TACE/ADAM17) is one of the candidate α-secretases. Initially identified as the protease responsible for pro-TNF-α cleavage, TACE's action of releasing TNF-α from its membrane-bound precursor is very similar to the processing of APP by α-secretase [11]. Buxbaum et al. [11] showed that knock-out of the TACE gene in primary embryonic fibroblasts almost completely abolished PKC-stimulated sAPPα production, but had little effect on basal sAPPα secretion. Furthermore, TACE was found to cleave a short synthetic peptide at the α-secretase site, indicating that the enzyme has the appropriate cleavage specificity [11].

TACE belongs to the ADAM (a disintegrin and metalloprotease) family, which is structurally related to the PIII class of snake venom metalloproteases (see Chapter 11). Members of this family contain a propeptide domain that must be removed to enable the enzyme to become active, a metalloprotease domain (which contains the catalytic activity), a disintegrin domain, a cysteine-rich domain and a transmembrane domain (Figure 3).

Owing to its structural similarity to TACE (Figure 3), ADAM10, another member of the ADAM family, has also been proposed to be an α-secretase. ADAM10 can cleave APP at the α-secretase site, and overexpression of ADAM10 increases sAPPα levels, which can be further increased by PKC [12]. In contrast to TACE, ADAM10 has both the basal and regulated components of α-secretase activity. A dominant-negative form of the protein inhibits both basal and phorbol ester stimulated α-secretase activity [12]. Furthermore, ADAM10 mRNA, unlike TACE, is co-localized with APP and a putative β-secretase candidate in the mouse brain [13].

A third putative α-secretase, PC7, belongs to a different family of proteases known as the proprotein convertases (PCs). This family of proteases is involved in proprotein processing (see Chapter 7). All members of this family are serine proteases that have a catalytic triad of residues at the active site (Figure 3). PC7 is expressed in a range of tissues and is localized to the *trans*-Golgi network. While overexpression of PC7 increases sAPPα production [10], it is still unclear whether PC7 is an authentic α-secretase. Recent evidence suggests that PC7 may proteolytically cleave the inactive pro-ADAM10 into the active form [14].

## β-Secretase

β-Secretase-mediated cleavage is the penultimate step in the production of Aβ. β-Secretase cleaves at two sites within the ectodomain of APP. The main cleavage site is located on the N-terminal side of $Asp^1$ of the Aβ sequence, with a minor cleavage site at $Glu^{11}$ (Figure 1). Cleavage at $Asp^1$ produces a

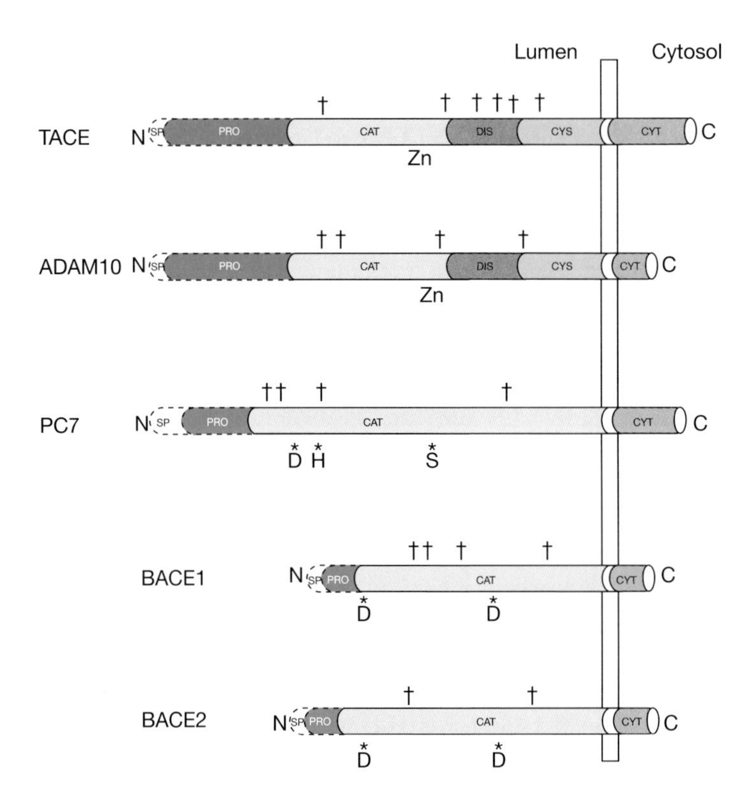


**Figure 3. Structural and functional domains of putative α- and β-secretases**
TACE and ADAM10 contain zinc-binding regions (Zn) within the ectodomain. PC7 has a catalytic triad, containing aspartate (D), histidine (H) and serine (S) residues. BACE1 and BACE2 contain aspartate residues, which form the active site. SP, signal peptide; PRO, proprotein sequence; CAT, catalytic domain; DIS, disintegrin domain; CYS, cysteine-rich domain; CYT, cytosolic domain. †, potential N-linked glycosylation sites; *, putative catalytic residues.

soluble C-terminally truncated fragment of APP that is eventually secreted (sAPPβ), and a membrane-bound fragment (C99). Subsequent cleavage of C99 by γ-secretase releases Aβ, which is also secreted (Figure 2). Most of the β-secretase appears to be localized in the Golgi apparatus [15] and astrocytes possess a low level of β-secretase activity. Thus, neurons are likely to be the major cell type contributing to Aβ production in the central nervous system.

In late 1999, four groups almost simultaneously reported the discovery of a protein that fulfilled all of the requirements of β-secretase [15–18]. This protease is known as BACE1 (β-site APP-cleaving enzyme 1), but has also been referred to as Asp2 and memapsin 2. BACE1 represents the first member of a new family of aspartic proteases. The protease is expressed in low levels in the periphery, but has a high expression level in the brain, particularly in neurons in the hippocampus, cortex and cerebellum [15,17], coinciding with the expected distribution of β-secretase. The gene encoding BACE1 is located on chromosome 11 and has an open reading frame of 501 amino acids. The gene encodes a type 1 integral membrane protein with several distinctive domains, including a catalytic domain, which contains two active-site regions containing the signature sequence of an aspartic protease (DT/SGT/S) (Figure 3). Consistent with its membership of this protease family, mutating the aspartate residues within the active site renders the protein inactive [15,16]. Despite conservation within the active-site region, BACE1 only has 30–37% identity with other known aspartic proteases. BACE1 is the first reported aspartic protease that contains a transmembrane domain, and also the first aspartic protease known to be insensitive to the protease inhibitor pepstatin.

Several lines of evidence demonstrate that BACE1 is a β-secretase. BACE1 can cleave full-length APP at both the major ($Asp^1$) and minor ($Glu^{11}$) β-secretase sites. The Swedish APP mutation, which increases β-secretase activity in cell culture, also increases the BACE1-catalysed cleavage at position 1 of the Aβ sequence. In cells that overexpress BACE1, there is increased production of all β-secretase products (Aβ, sAPPβ and C99) [15–17]. Furthermore, BACE1 antisense oligonucleotides inhibit β-secretase activity [15]. There is almost total loss of Aβ compared to controls in BACE1 knock-out mice [19], and cortical neurons cultured from BACE1 knockout embryos do not produce any β-secretase-derived fragments of APP (C99, Aβ1–40/42 and Aβ11–40/42), although there is an increase in α-secretase-generated products [20]. Taken together, these findings strongly argue that BACE1 is the major β-secretase in neurons.

BACE1 undergoes several post-translational modifications. In the Golgi apparatus, phosphorylation of the C-terminus occurs and the pro-peptide domain is cleaved, producing the mature form of BACE1. A proprotein convertase (furin) cleaves off the propeptide domain of BACE1. Unlike most aspartic proteases, BACE1 does not appear to require propeptide cleavage to become active; pro-BACE1 does not significantly differ in activity from mature BACE1.

Another transmembrane aspartic protease has been named BACE2 owing to its high identity with BACE1 (Figure 3). Although BACE2 can cleave APP at the β-secretase sites ($Asp^1$ and $Glu^{11}$), it is more efficient at cleaving Aβ internally at $Phe^{19}$ and $Phe^{20}$ [21]. Furthermore, BACE2 is found primarily in the periphery, with negligible levels in the brain [21]. Therefore, although BACE2 may have some β-secretase ability, it is unlikely (owing to its cleavage sites and localization) to contribute significantly to Aβ formation in the brain. Nonetheless, selective inhibitors of BACE1, which do not block BACE2, may need to be identified if the inhibition of BACE2 has toxic consequences.

## γ-Secretase

Cleavage of the β-secretase product C99 by γ-secretase is the final step in the production of Aβ. γ-Secretase can cut C99 at a number of sites (Figure 1), producing a variety of Aβ species with different C-termini. Aβ40 and Aβ42 (long Aβ) are major products of γ-secretase activity.

The location of the γ-secretase cleavage site is unusual as it resides within the transmembrane region. Several transmembrane proteins are known to be cleaved within or close to their membrane-spanning domain. This phenomenon, known as regulated intramembrane proteolysis, is discussed in more detail elsewhere in this volume (Chapter 12). At present, the mechanism by which regulated intramembrane proteolysis occurs is unknown. Cleavage by proteases is a hydrolytic event that requires the presence of water. It is unclear how water can get to the active site in the hydrophobic environment of the membrane. Cleavage may occur within the membrane by some as yet undefined mechanism. Alternatively, proteolytic cleavage may occur after the polypeptide has been removed from the lipid bilayer.

Although a number of proteins have been proposed to have γ-secretase activity, thus far, none has convincingly demonstrated the characteristics required for γ-secretase. Integral membrane proteins known as presenilins are γ-secretase candidates. The genes for presenilin-1 (PS1) and presenilin-2 (PS2) are located on chromosomes 14 and 1 respectively, and encode large proteins with 67% identity [22]. The proposed structure of the presenilins has eight transmembrane regions with a large cytosolic loop (Figure 4). The presenilins can be cleaved by a protease known as presenilinase, producing N- and C-terminal fragments of 35 and 20 kDa respectively, as well as by caspase at an adjacent site in the loop. The fragments generated by presenilinase remain associated after cleavage.

Mutations in the presenilin genes can lead to early-onset AD, and are responsible for more familial cases of AD than mutations of the APP gene. At the time of writing, 85 mutations in the PS1 gene have been identified and five are known within the PS2 gene [22]. The close association between presenilin and γ-secretase is suggested by the demonstration that knock-out of the presenilin genes results in knock-out of γ-secretase activity. The absence of presenilins eliminates Aβ production, with accumulation of α- and β-secretase-cleaved products. Furthermore, mutations in presenilins affect the γ-secretase cleavage site, increasing the amount of Aβ42 produced. Two aspartate residues, located in the sixth and seventh transmembrane domains, that are critical for presenilin function are predicted to align in the membrane and have been proposed to give rise to a catalytic cleavage similar to that of the aspartic proteases [22].

However, the relationship between presenilins and γ-secretase remains unclear. Overexpression of presenilin does not lead to increased γ-secretase activity, nor has either of the presenilins been shown to possess protease activity. Furthermore, the subcellular localization of PS1 does not always appear to correspond with the localization of γ-secretase activity [23].

Although presenilins may not be identical with γ-secretase, they may form part of a high molecular mass complex that contains the γ-secretase. Immunoprecipitation of presenilins leads to the co-immunoprecipitation of

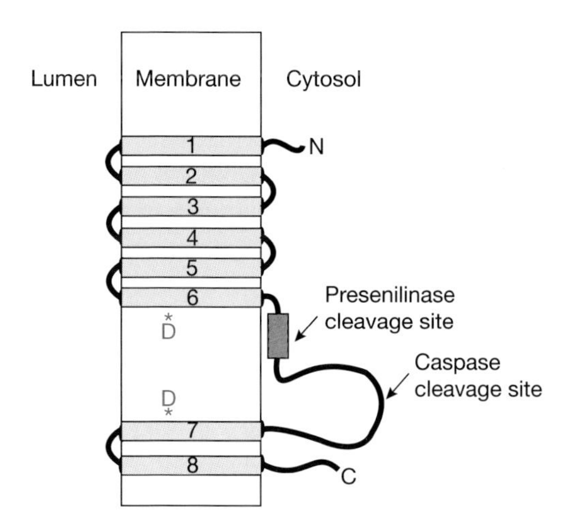


**Figure 4. Structural domains of presenilins**
Presenilins have nine hydrophobic regions, eight of which are predicted to span the membrane. Presenilinase and caspase cleavage sites occur in the large cytosolic loop. Two aspartate residues (D) are predicted to align within the membrane. * Putative catalytic aspartate residues.

γ-secretase activity. Furthermore, photoactivatable γ-secretase inhibitors bind to presenilin heterodimers [24]. Recently, a type 1 transmembrane protein known as nicastrin, which binds to full-length APP as well as α-secretase- and β-secretase-cleaved APP, and forms a high molecular mass complex containing presenilin, was identified [25]. The identification of other proteins in this complex is an active area of investigation at present.

## Secretases as therapeutic targets for AD

As the proteolytic processing of APP to Aβ is central to the pathogenesis of the disease, drugs that inhibit amyloidogenic processing of APP are potential therapeutic agents. For example, the upregulation of α-secretase activity may channel APP processing down a non-amyloidogenic pathway, thereby decreasing Aβ production. A number of agents increase α-secretase activity activating PKC, e.g. muscarinic receptor stimulation [5]. Therefore, cholinesterase inhibitors that boost levels of acetylcholine may increase non-amyloidogenic processing of APP, by decreasing Aβ production and slowing the progression of AD. However, studies using long-term anti-cholinesterase treatment of AD patients have provided little clear evidence of a definitive slowing of the progression of the disease. Similarly, there is as yet no evidence that muscarinic receptor agonists improve cognition in AD.

At present, the most attractive therapeutic strategy is based upon the direct inhibition of the amyloidogenic pathway. One approach is to inhibit β-secretase. The crystal structure of BACE1 has been determined recently, and indicates that the active site is quite distinctive, being more hydrophobic and open than that of other aspartic proteases [26]. This is an advantage for selective inhibitor design. Several compounds that inhibit Aβ production by reducing cleavage by β-secretase have been identified. One of the most potent of these is a tripeptide aldehyde that contains reduced amide isoteres that have been used previously for aspartic protease inhibition. As BACE1 gene knock-out mice are viable, with no major physical or cognitive deficits [19], this suggests that inhibition of BACE1 may be an effective therapeutic strategy with minimal toxic consequences.

The alternative approach is to inhibit γ-secretase. This should also decrease Aβ production. A number of effective γ-secretase inhibitors have been reported; one very potent inhibitor is a compound with an hydroxyethylene dipeptide isotere, which acts as a transition state analogue mimic at the catalytic site of an aspartic protease [27]. However, the efficacy of γ-secretase inhibitors for AD therapy is uncertain, as there may be other normal physiological substrates for γ-secretase. For example, γ-secretase can also cleave a protein called Notch, which is similar to APP, and is essential for a number of cellular functions, such as cell fate and differentiation [28]. The intramembraneous cleavage of Notch releases the intracellular domain, which translocates to the nucleus and activates the CSL family of transcription factors. Notch function relies on this cleavage; its absence causes a lethal phenotype in embryos. The protease required for Notch processing is strikingly similar to γ-secretase (see Chapter 12), and the majority of inhibitors are unable to distinguish between the two proteolytic activities. However, the recent development of γ-secretase inhibitors that have greater selectivity for APP processing than for Notch processing [29] indicates that, despite their similarity, the γ-secretase-like proteases cleaving APP and Notch may be pharmacologically different.

## Conclusions

There is very good evidence to suggest that the accumulation of Aβ is the underlying cause of AD, and that inhibition of Aβ production may prevent the progression of the disease. Much progress has been made in understanding the proteases responsible for the cleavage of Aβ from APP. α-Secretase, the protease responsible for the majority of APP processing, may be associated with one or more members of the ADAM family of metalloproteases. β-Secretase has been identified as an aspartic protease (BACE1) and γ-secretase has yet to be identified definitively. At the time of writing, several secretase inhibitors are being tested in phase 1 clinical trials. The results of these human trials will eventually indicate whether secretase inhibitors are effective in the treatment or prevention of AD, but they may also provide the first direct proof of the amyloid hypothesis of AD.

## Summary


- *The accumulation of Aβ is believed to be the underlying cause of the neurodegeneration seen in AD patients. This is known as the amyloid hypothesis.*
- *Aβ is formed by the proteolytic cleavage of its precursor, APP.*
- *Cleavage of APP by β- and γ-secretases produces the N- and C-termini of Aβ respectively. Cleavage by α-secretase destroys the Aβ sequence.*
- *A number of α-secretase candidates have been proposed, and it appears likely that more than one protease is responsible for cleavage at the α-secretase site.*
- *A recently identified protease named BACE1 is responsible for the majority of β-secretase activity in the central nervous system.*
- *There are a number of sites at which γ-secretase can cleave, all of which are located in the transmembrane domain of APP. Although the enzyme(s) responsible for these cleavages has not been identified, the presenilins are likely to play a major role.*
- *The inhibition of Aβ production is an attractive therapeutic strategy. This may be achieved by the selective inhibition of β-secretase and/or γ-secretase.*


## References


1. Probst, A., Langui, D. & Ulrich, J. (1991) Alzheimer's disease: a description of the structural lesions. *Brain Pathol.* **1**, 229–239
2. Masters, C.L., Simms, G., Weinman, N.A., Multhaup, G., McDonald, B.L. & Beyreuther, K. (1985) Amyloid plaque core protein in Alzheimer disease and Down syndrome. *Proc. Natl. Acad. Sci. U.S.A.* **82**, 4245–4249
3. Kang, J., Lemaire, H.-G., Unterbeck, A., Salbaum, J.M., Masters, C.L., Grzeschik, K.H., Multhaup, G., Beyreuther, K. & Müller-Hill, B. (1987) The precursor of Alzheimer's disease A4 protein resembles a cell surface receptor. *Nature (London)* **325**, 733–736
4. Gandy, S. & Petanceska, S. (2000) Regulation of Alzheimer β-amyloid precursor trafficking and metabolism. *Biochim. Biophys. Acta* **1502**, 44–52
5. Nunan, J. & Small, D.H. (2000) Regulation of APP cleavage by α-, β- and γ-secretases. *FEBS Lett.* **483**, 6–10
6. Small, D.H. & McLean, C.A. (1999) Alzheimer's disease and the amyloid β protein: what is the role of amyloid? *J. Neurochem.* **73**, 443–449
7. Sheuner, D., Eckman, C., Jensen, M., Song, X., Citron, M., Suzuki, N., Bird, T.D., Hardy, J., Hutton, M., Kukull, W. et al. (1996) Secreted amyloid β-protein similar to that in the senile plaques of Alzheimer's disease is increased in vivo by the presenilin 1 and 2 and APP mutations linked to familial Alzheimer's disease. *Nat. Med.* **2**, 864–870
8. Small, D.H., Mok, S.S. & Bornstein, J.C. (2001) Alzheimer's disease and Aβ toxicity: from top to bottom. *Nat. Rev.* **2**, 595–598
9. Schenk, D., Barbour, R., Dunn, W., Gordon, G., Grajeda, H., Guido, T., Hu, K., Huang, J., Johnson-Wood, K., Khan, K. et al. (1999) Immunization with amyloid-β attenuates Alzheimer-disease-like pathology in the PDAPP mouse. *Nature (London)* **400**, 173–177

10. Lopez-Perez, E., Zhang, Y., Frank, S.J., Creemers, J., Seidah, N. & Checler, F. (2001) Constitutive α-secretase cleavage of the β-amyloid precursor protein in the furin-deficient LoVo cell line: involvement of the pro-hormone convertase 7 and the disintegrin metalloprotease ADAM10. *J. Neurochem.* **76**, 1532–1539
11. Buxbaum, J.D., Liu, K-N., Luo, Y., Slack, J.L., Stocking, K.L., Peschon, J.J., Johnson, R.S., Castner, B.J., Cerretti, D.P. & Black, R.A. (1998) Evidence that tumor necrosis factor α converting enzyme is involved in regulated α-secretase cleavage of the Alzheimer amyloid protein precursor. *J. Biol. Chem.* **273**, 27765–27767
12. Lammich, S., Kojro, E., Postina, R., Gilbert, S., Pfeiffer, R., Jasionowski, M., Haass, C. & Fahrenholz, F. (1999) Constitutive and regulated α-secretase cleavage of Alzheimer's amyloid precursor protein by a disintegrin metalloprotease. *Proc. Natl. Acad. Sci. U.S.A.* **96**, 3922–3927
13. Marcinkiewicz, M. & Seidah, N.G. (2000) Coordinated expression of β-amyloid precursor protein and the putative β-secretase BACE and α-secretase ADAM10 in mouse and human brain. *J. Neurochem.* **75**, 2133–2143
14. Gilbert, S., Kojro, E., Anders, A., Postina, R. & Fahrenholz, F. (2001) Regulation of ADAM10 alpha-secretase activity. *J. Neurochem.* **78** (Suppl. 1), 9
15. Vassar, R., Bennett, B.D., Babu-Khan, S., Kahn, S., Mendiaz, E.A., Denis, P., Teplow, D.B., Ross, S., Amarante, P., Loeloff, R. et al. (1999) β-Secretase cleavage of Alzheimer's amyloid precursor protein by the transmembrane aspartic protease BACE. *Science* **286**, 735–741
16. Hussain, I., Powell, D., Howlett, D.R., Tew, D.G., Meek, T.D., Chapman, C., Gloger, I.S., Murphy, K.E., Southan, C.D., Ryan, D.M. et al. (1999) Identification of a novel aspartic protease (Asp2) as β-secretase. *Mol. Cell. Neurosci.* **14**, 419–427
17. Sinha, S., Anderson, J.P., Barbour, R., Basi, G.S., Caccavello, R., Davis, D., Doan, M., Dovey, H.F., Frigon, N., Hong, J. et al. (1999) Purification and cloning of amyloid precursor protein β-secretase from human brain. *Nature (London)* **402**, 537–540
18. Yan, R., Bienkowski, M.J., Shuck, M.E., Miao, H., Tory, M.C., Pauley, A.M., Brashler, J.R., Stratman, N.C., Mathews, W.R., Buhl, A.E. et al. (1999) Membrane-anchored aspartyl protease with Alzheimer's disease β-secretase activity. *Nature (London)* **402**, 533–537
19. Luo, Y., Bolon, B., Kahn, S., Bennett, B.D., Babu-Khan, S., Denis, P., Fan, W., Kha, H., Zhang, J., Gong, Y. et al. (2001) Mice deficient in BACE1, the Alzheimer's β-secretase, have normal phenotype and abolished β-amyloid generation. *Nat. Neurosci.* **4**, 231–232
20. Cai, H., Wang, Y., McCarthy, D., Wen, H., Borchelt, D.R., Price, D.L. & Wong, P.C. (2001) BACE1 is the major β-secretase for generation of Aβ peptides by neurons. *Nat. Neurosci.* **4**, 233–234
21. Farzan, M., Schnitzler, C.E., Vasilieva, N., Leung, D. & Choe, H. (2000) BACE2, a β-secretase homolog, cleaves at the β site and within the amyloid-β region of the amyloid-β precursor protein. *Proc. Natl. Acad. Sci. U.S.A.* **97**, 9712–9717
22. Wolfe, M.S. (2001) Presenilin and γ-secretase: structure meets function. *J. Neurochem.* **76**, 1615–1620
23. Cupers, P., Bentahir, M., Craessaerts, K., Orlans, I., Vanderstichele, H., Saftig, P., DeStrooper, B. & Annaert, W. (2001) The discrepancy between presenilin subcellular localization and γ-secretase processing of amyloid precursor protein. *J. Cell Biol.* **154**, 731–740
24. Li, Y.-M., Xu, M., Lai, M.-T., Huang, Q., Castro, J.L., DiMuzio-Mower, J., Harrison, T., Lellis, C., Nadin, A., Neduvelil, J.G. et al. (2000) Photoactivated γ-secretase inhibitors directed to the active site covalently label presenilin 1. *Nature (London)* **405**, 689–694
25. Yu, G., Nishimura, M., Arawaka, S., Levitan, D., Zhang, L., Tandon, A., Song, Y.-Q., Rogaev, E., Chen, F., Kawarai, T. et al. (2000) Nicastrin modulates presenilin-mediated notch/glp-1 signal transduction and βAPP processing. *Nature (London)* **407**, 48–54
26. Hong, L., Koelsch, G., Lin, X., Wu, S., Terzyan, S., Ghosh, A.K., Zhang, X.C. & Tang, J. (2000) Structure of the protease domain of memapsin2 (β-secretase) complexed with inhibitor. *Science* **290**, 150–153

27. Shearman, M.S., Beher, D., Clarke, E.E., Lewis, H.D., Harrison, T., Hunt, P., Nadin, A., Smith, A.L., Stevenson, G. & Castro, J.L. (2000) L-685,458, an aspartyl protease transition state mimic, is a potent inhibitor of amyloid β-protein precursor γ-secretase activity. *Biochemistry* **39**, 8698–8704
28. Struhl, G. & Adachi, A. (1998) Nuclear access and action of notch in vivo. *Cell* **93**, 649–660
29. Petit, A., Bihel, F., Alvès da Costa, C., Pourquié, O., Checler, F. & Kraus, J.-L. (2001) New protease inhibitors prevent γ-secretase-mediated production of Aβ40/42 without affecting Notch cleavage. *Nat. Cell Biol.* **3**, 507–511

# 5

# The ubiquitin–proteasome pathway of intracellular proteolysis

Fergus J. Doherty, Simon Dawson and R. John Mayer[1]

*Laboratory of Intracellular Proteolysis, School of Biomedical Sciences, University of Nottingham Medical School, Queen's Medical Centre, Nottingham NG7 2UH, U.K.*

## Abstract

Intracellular proteins are targeted for degradation by the covalent attachment of chains of the small protein ubiquitin; a process known as ubiquitylation. Many proteins are phosphorylated prior to ubiquitylation, and therefore ubiquitylation and degradation of these proteins is regulated by kinase activity and signalling cascades. Many ubiquitylated proteins are degraded by the 26 S proteasome complex, which is found in the cytosol and nucleus. The 26 S proteasome consists of a 20 S core with proteolytic activity and 18 S regulatory complexes containing ATPases and ubiquitin-chain-binding proteins. Proteins degraded by the ubiquitin–proteasome pathway include cyclins and other regulators of the cell cycle, and transcription factors. Abnormal polypeptides are also degraded by the ubiquitin pathway, including abnormal polypeptides in the endoplasmic reticulum, which are translocated back out of the endoplasmic reticulum prior to ubiquitylation and degradation by the proteasome. The ubiquitin–proteasome pathway is implicated in numerous diseases including cancer and neurodegenerative diseases.

[1]*To whom correspondence should be addressed (e-mail: john.mayer@nottingham.ac.uk).*

## Introduction

The proteasome is a large multi-subunit protease found in eukaryotic cells and archaebacteria, which appears to be responsible for the majority of the degradation of intracellular proteins. Most proteins are targeted for destruction by the proteasome by being covalently tagged with the small protein ubiquitin. The attachment of ubiquitin to target proteins, ubiquitylation, is itself a complex multi-step process and is responsible for much of the selectivity of proteasome-mediated proteolysis. The ubiquitin–proteasome pathway is implicated in numerous cell processes through the regulated degradation of key proteins, often in response to signalling events. In addition, the ubiquitin–proteasome pathway is involved in the removal of damaged or abnormal proteins. Given the important role of this pathway, it is not surprising to find that it is involved in several disease processes.

## Ubiquitin: marking proteins for destruction

The crystal structure of ubiquitin [1], which is only found in eukaryotes, reveals a small compact 'peardrop-like' protein with stretches of interspersed α-helices and β-sheets (Figure 1a). The 'business end' of the protein extends into the aqueous space and can be used in the formation of an isopeptide bond (ubiquitylation) between the carboxyl group of the C-terminal glycine and the side-chain amino group of a lysine residue in target proteins, or indeed in another copy of ubiquitin (Figure 1b). Ubiquitin contains seven lysine residues, many of which can form isopeptide bonds to give rise to families of distinct ubiquitin chains. Ubiquitin $Lys^{48}$-linked chains (between the side-

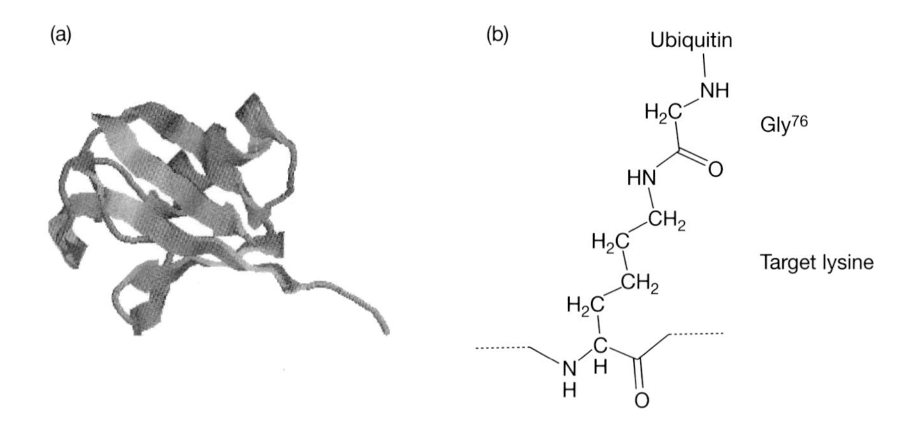


**Figure 1. Ubiquitin and the ubiquitin–protein isopeptide bond**
(a) Ubiquitin has a compact globular shape containing β-sheets and α-helices. The C-terminus extends into the aqueous space. This model is generated from 1ubq.pdb. (b) The isopeptide bond between the C-terminal carboxyl group of ubiquitin ($Gly^{76}$) and a lysine side chain amino group of a target protein.

chain amino group of $Lys^{48}$ of one ubiquitin and the C-terminal group of another ubiquitin and with at least four ubiquitins in the chain) conjugated to target proteins are signals for degradation of the isopeptide-linked target protein. Chains of ubiquitins that are linked through other lysine residues may be involved in other cellular pathways.

A number of ubiquitin-like proteins (ubiquitin paralogues) have been discovered, including SUMO (small ubiquitin-like modifier). Sumoylation of proteins may prevent their degradation by preventing ubiquitylation [2].

## Ubiquitylation and de-ubiquitylation

Ubiquitylation of target proteins is a multi-step process that starts with ubiquitin activation. The ubiquitin-activating enzyme (E1 in Figure 2) promotes the formation of ubiquitin–adenylate. The activated ubiquitin then forms a thiolester with the activating enzyme, with the release of AMP. Ubiquitin is then transferred to one of a family of conjugating enzymes (E2) in the form of a thiolester [3]. Another class of enzymes, E3s or ubiquitin–protein ligase, is necessary for the transfer of the ubiquitin to the target protein, with the associated formation of an isopeptide bond.

E3 proteins or protein complexes recognize target proteins and ubiquitin-charged E2 proteins, and bring them together for ubiquitylation of the target protein. Therefore, it appears that the specificity of the ubiquitin-mediated proteolytic machinery depends upon the E3 proteins. Ligases must recognize features of the target protein, and a number of these degradation signals, or degrons, have been identified. Many proteins must be phosphorylated prior to ubiquitylation and, therefore, the phosphorylated residue forms part of the degron.

### De-ubiquitylation — an important regulatory mechanism

There are many de-ubiquitylating enzymes (DUBs), which implies that de-ubiquitylation may have an important regulatory role in protein degradation [4]. There are two classes of DUBs. Ubiquitin C-terminal hydrolases cleave ubiquitin from the fusion proteins that are the products of ubiquitin genes, thereby releasing ubiquitin monomers and the unrelated C-terminal extension proteins. Ubiquitin-specific proteases (UBPs) cleave ubiquitin from multi-ubiquitin chains. The activities of UBPs may, therefore, function by saving proteins from degradation. The proteasome contains a UBP activity, which trims long multi-ubiquitin chains on target proteins. If one considers the large numbers of DUBs (19 in *Saccharomyces cerevisiae*), it may be that they have an important role in regulating intracellular proteolysis. Several DUBs are known to be involved in specific cell processes and some are involved in disease.

## The big mean proteolytic machine: the proteasome

Once a target protein is marked by multi-ubiquitylation, it appears to have a short half-life in the cell as it is degraded rapidly by the proteasome. The

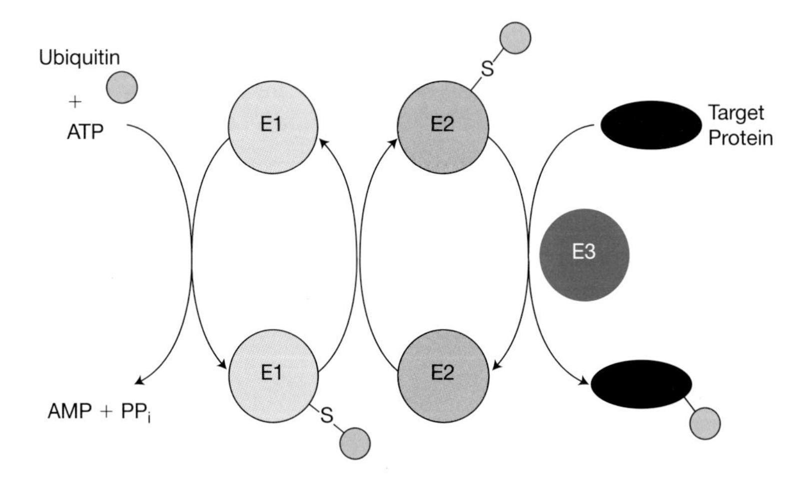


**Figure 2. Ubiquitin activation and conjugation**
EI (or UbA in yeast) is the ubiquitin-activating enzyme, which initially forms a ubiquitin–adenylate complex (not shown) and then forms a thiol ester between the carboxyl group of the C-terminal glycine of ubiquitin and a cysteine residue of EI. The ubiquitin is then transferred to one of a large family of ubiquitin-conjugating enzymes (E2; Ubc in yeast), the ubiquitin forming a thiolester as before with a cysteine residue of the enzyme. Finally, ubiquitin is transferred to the target protein, forming an isopeptide bond between the C-terminus of ubiquitin and an $\epsilon$-amino group in the side chain of a lysine residue of the target protein. The transfer of ubiquitin to the target can occur directly from a charged E2, with E3 (ubiquitin ligase) recognizing the target and bringing it and charged E2 together to form a complex. Alternatively, some E3s act as ubiquitin acceptors and ubiquitin is transferred to the E3 to form a thiolester before being transferred to the target protein, which is also recognized by this class (HECT; homologous with E6-associated protein carboxy terminus) of E3/ligases. Further copies of ubiquitin can then be conjugated to ubiquitin attached to the protein to form a multi-ubiquitin chain. Ubiquitin–ubiquitin isopeptide bonds are between the C-terminal carboxyl group of the distal ubiquitin and the $\epsilon$-amino group of $Lys^{48}$ or $Lys^{63}$ of the proximal ubiquitin. $Lys^{48}$-based multi-ubiquitin chains of at least four ubiquitin molecules are necessary to target proteins to the 26 S proteasome.

multi-subunit proteasome is a compartmentalized protease: the active sites ($\beta$-subunits) are within the 20 S proteolytic core which consists of 28 subunits ($\alpha_7$, $\beta_7$, $\beta_7$, $\alpha_7$) [5] (Figure 3). An intracellular protease should not expose catalytic sites to the cellular milieu, as this would result in the inappropriate degradation of proteins.

The proteasome is a threonine protease, with the threonine residue at the N-terminus of at least three of the $\beta$-subunits that are involved in nucleophilic attack [6,7]. This threonine can be replaced by serine with no loss of activity [7].

Ubiquitylated proteins cannot gain access to the central 20 S cavity that contains the catalytic sites: in yeast, and presumably higher organisms, the ends of the 20 S cylinders are closed by the N-termini of the $\alpha$-subunits [8]. These termini must be opened to enable target proteins to gain access to the catalytic sites. Furthermore, the opened pore or hole is still too small for globular proteins. Therefore, the heptameric $\alpha$-subunit rings at each end of the 20 S cylinder can bind to a 19 S regulator, giving rise to the 26 S proteasome that is

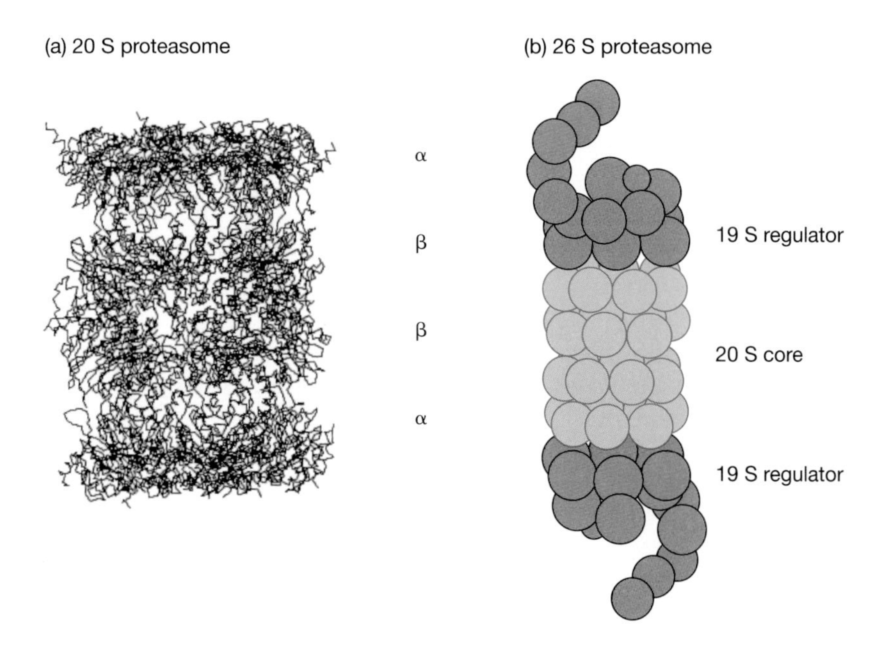


**Figure 3. The proteasome**
(a) A backbone model of the 20 S proteasome of yeast generated from crystallographic data. The α and β rings are indicated. (b) A representation of 26 S proteasome based on electron micrographs.

responsible for degrading ubiquitylated proteins [9]. Each 19 S regulator consists of a 'base', which binds to each α-subunit heptameric ring, and a 'lid'. The base contains six non-identical related ATPases, which are presumed to form a hexameric complex [9] as in prokaryotic ATP-dependent proteases (Figure 3). The ATPases are molecular chaperones, which, like the ATPase subunits of prokaryotic protease complexes, unfold target proteins. The lid contains proteins such as S5a (otherwise known as Rpn10 and Mcb1), which appears to be involved in the recognition of multi-ubiquitylated proteins [10]. The ATPases may also be involved directly or indirectly in recognizing multi-ubiquitylated proteins. There are protein substrates that are degraded by the 26 S proteasome that do not have to be ubiquitylated. An example is ornithine decarboxylase, which, following association with a protein known as antizyme [11], is degraded by the 26 S proteasome. Other substrates, such as the cyclin-dependent kinase inhibitor, p21, can be degraded by the 20 S core proteasome lacking the 26 S regulator, after binding to a 20 S subunit [12].

Other regulators of the 20 S proteasome exist. The most studied is the 11 S regulator or PA28. This heptameric assembly can attach to one end of a 20 S cylinder with a 19 S regulator at the other end. The binding of 11 S to the 20 S core appears to facilitate the cleavage of substrate proteins to give rise to peptides which are more suited to the binding of MHC Class I (MHC I) molecules [13]. Generation of peptide antigens is one of the key roles of the proteasome.

As might be anticipated, proteins which are not part of these complexes can interact with the complexes to modulate the proteolytic activity. Proteins that bind to individual ATPases in the putative hexameric ring of the 19 S base include physiological regulators like gankyrin (which targets retinoblastoma protein for degradation by the proteasome) [14] and proteins from pathogens, e.g. human papilloma virus E7 protein, which also targets pRb for degradation [15]. It is likely that many other proteins that interact with different subunits of both the proteasomal 19 S regulator and the 20 S core will be discovered in due course.

## Intracellular pathways controlled by the ubiquitin–proteasome system

Numerous cellular processes involve regulated proteolysis of key components by the ubiquitin–proteasome system. Phosphorylation events often trigger degradation of a protein by the ubiquitin–proteasome pathway. It is not possible in this review to cover many examples of protein degradation by this system; a few are included here.

### Cell cycle

Each step in the cell cycle is controlled by cyclin-dependent protein kinases (Cdks) and, in turn, by the degradation of cyclins and kinase inhibitors by the ubiquitin–proteasome pathway. The accumulation of a cyclin, and therefore the activation of a Cdk, will promote entry into a particular stage of the cell cycle. These factors will also inhibit entry into the next stage of the cycle until, for instance, the cyclin is degraded. This ensures that the cell cycle is composed of distinct stages. Mutations in kinases, kinase inhibitors and components of ubiquitin protein ligases are found in cancer cells. The E3 proteins involved in these degradative events are the multi-component SCF (Skip–cullin–F-box) type ligase and phosphorylation of the target protein often triggers ubiquitylation. A related complex, the anaphase-promoting complex (APC; cyclosome), which ubiquitylates mitotic cyclins, contains a RING-finger protein (Apc11), an E2 and a cullin (Cdc20/Hct1), but Skp1 is replaced by multiple APC subunits [16]. The APC is involved in the degradation of Pds1/Cut1, factors involved in the regulation of daughter chromatid separation. Regulation of the cell cycle is highly complex, with many components of the cycle control mechanism subjected to rapid degradation by the ubiquitin–proteasome pathway (Figure 4).

### Transcription

Transcription factors that control the expression of eukaryotic genes are essential but potentially lethal molecules. Aberrant transcription is a very significant feature of cancer. Transcription factors are very short-lived and are degraded quickly by the ubiquitin pathway. One of the best studied examples is slightly different in that an inhibitor of a transcription factor is degraded, thus

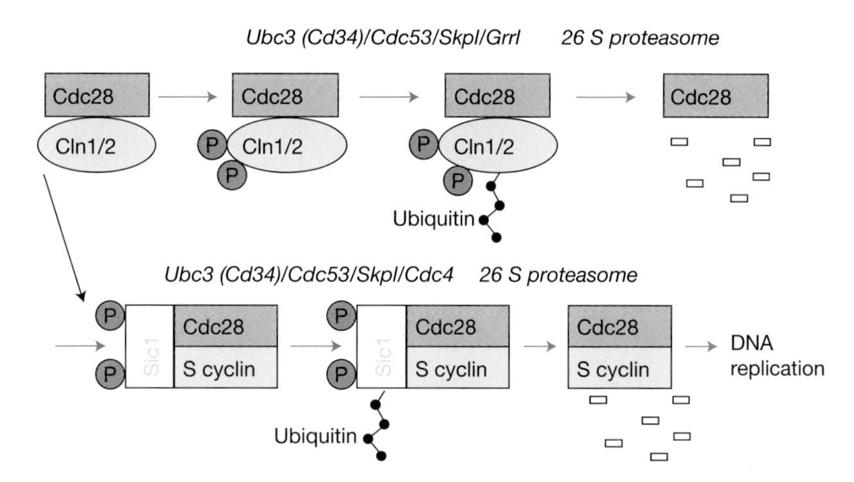


**Figure 4. Ubiquitin-mediated proteolysis and $G_1$/S phase transition in *Saccharomyces cerevisiae***

At the start of $G_1$, constitutive Cdk (Cdc28) activates the expression of cyclins 1 and 2 (Cln1/2). Cln1/2 further activate Cdc28, which results in even further Cln1/2 expression. At a certain threshold of Cln1/2–Cdc28, phosphorylation of the Cdk inhibitor Sic1 occurs, which activates S-phase cyclin (Clb)–Cdc28 complexes, which in turn activate DNA synthesis. In addition, the increased levels of Cln1/2–Cdc28 complexes lead to phosphorylation of Cln1/2, triggering their ubiquitylation and degradation, and the cell can enter S-phase. Phosphorylation on Cln1 and Cln2 occurs in PEST regions (rich in proline, glutamate, serine and threonine) and acts as a signal for their ubiquitylation.

releasing the transcription factor to cause gene expression. The transcription factor is nuclear factor κB (NFκB), which regulates the expression of many genes involved in inflammation and immunity. The inhibitor of NFκB (IκB), binds to NFκB and prevents the transcription factor from entering the nucleus and, therefore, binding to DNA. Once again, phosphorylation of the target protein, IκB, triggers ubiquitylation by an SCF ligase. IκB is phosphorylated by IκB kinase, which in turn is activated by the kinase tumour necrosis factor-activated kinase 1 (TAK1), which is activated by association with the self-ubiquitylated protein tumour necrosis factor-associated factor 6 (TRAF6) [17]. The multi-ubiquitin chain formed on TRAF6 is formed between $Lys^{63}$ residues of ubiquitin, rather than the $Lys^{48}$ residues that lead to degradation by the proteasome (Figure 5). This is an example of ubiquitylation as a regulatory covalent modification with a function distinct from proteolysis. Interestingly, a form of NFκB where the IκB inhibitor is incorporated into a NFκB precursor is partially degraded following phosphorylation and ubiquitylation, to generate active transcription factor. This is a situation where ubiquitylation does not lead to complete degradation of the ubiquitylated protein [18], another example being the release of a membrane-bound transcription factor resulting from limited proteolysis following ubiquitylation [19].

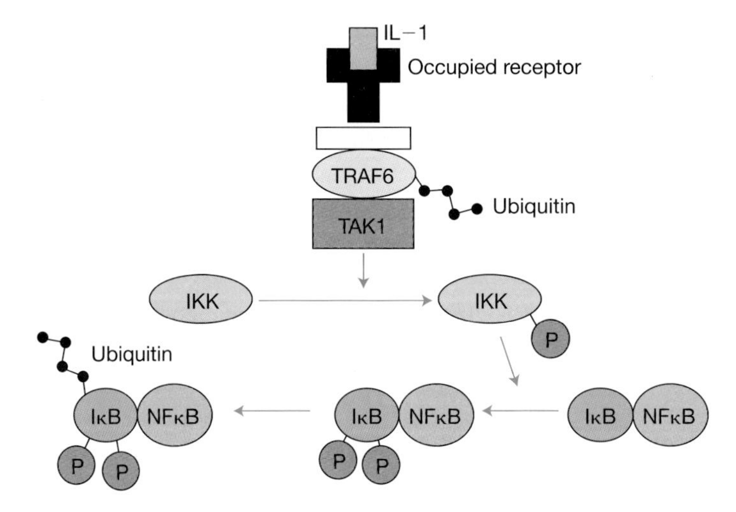


**Figure 5 Ubiquitylation and the regulation of the transcription factor NF-κB**
Binding of the cytokine interleukin-1 (IL-1) to its receptor at the cell surface leads to recruitment of adaptor proteins and TRAF6 to the cytoplasmic tail of the receptor. TRAF6 appears to auto-ubiquitylate. Ubiquitylated TRAF6 activates TAK1, a kinase that phosphorylates and activates IκB kinase (IKK). IKK in turn phosphorylates IκB, an inhibitor of the transcription factor NFκB. Phosphorylation of IκB leads to its recognition and ubiquitylation by an SCF ligase containing βTRCP (β-transducin repeats-containing protein) as the F-box protein. Ubiquitylation of IκB leads to its degradation by the 26 S proteasome, thereby freeing NFκB to enter the nucleus and activate genes involved in the inflammatory response. IκB is protected from ubiquitylation prior to phosphorylation by modification of the lysine acceptor sites with the ubiquitin-like protein SUMO (not shown). SUMO must be removed prior to IκB degradation.

## Protein folding versus degradation

Folding of newly synthesized polypeptides in the cytosol is often helped by molecular chaperones, e.g. Hsp70 (heat-shock protein 70). But what happens if a protein cannot fold or misfolds? The co-chaperone CHIP (C-terminal Hsp70-interacting protein) contains a modified RING-finger (the U-box) and promotes the ubiquitylation of misfolded proteins, which are then degraded by the proteasome [20]. The term 'defective ribosomal products' (DRIPs) has been coined for misfolded newly synthesized proteins. It is likely that repeated binding of a partially folded polypeptide with exposed hydrophobic residues to a Hsc70 (heat-shock cognate 70 stress protein) chaperone family member increases the likelihood of ubiquitylation by CHIP. Polypeptides that fold quickly may escape this fate. There is increasing evidence that a substantial fraction of all newly synthesized proteins are degraded to produce antigenic peptides for MHC I presentation (see later). Chaperones and CHIPs may be needed for this process. It is suggested, therefore, that CHIP is involved in 'protein triage' decisions in the cell, favouring degradation of target proteins over chaperone-mediated folding processes.

## Endoplasmic reticulum (ER)-associated degradation system

All membrane-associated and secreted proteins are synthesized on membrane-bound ribosomes and pass into the lumen of the ER. The lumen of the ER is, therefore, a very 'busy' place, filled with chaperones and enzymes involved in protein folding and post-translational modifications, e.g. glycosylation.

What happens to misfolded and misglycosylated proteins in the ER? Remarkably, they are extracted from the ER through the same channel by which they enter the lumen (the translocon), ubiquitylated and degraded by the proteasome (Figure 6). Both soluble lumenal ER proteins, e.g. a mutant carboxypeptidase in yeast, and ER membrane proteins, e.g. cystic fibrosis transmembrane conductance regulator [21], are degraded after extraction from the ER. Molecular chaperones interact with the misfolded proteins, and prevent them from aggregating in the ER and permit their dislocation to the cytosol for ubiquitylation and degradation. The presence of abnormal proteins in the ER triggers the unfolded protein response, which upregulates a number of genes, including some of those coding for components of the ubiquitin–proteasome pathway [22]. It would not be surprising if viruses did not exploit the ER degradation pathway. Cytomegalovirus and HIV have proteins that bind to MHC I molecules in the ER and force their ejection from the membrane, which is followed by their degradation by the ubiquitin pathway in order to reduce the MHC I immune response (see later) to viral proteins [23].

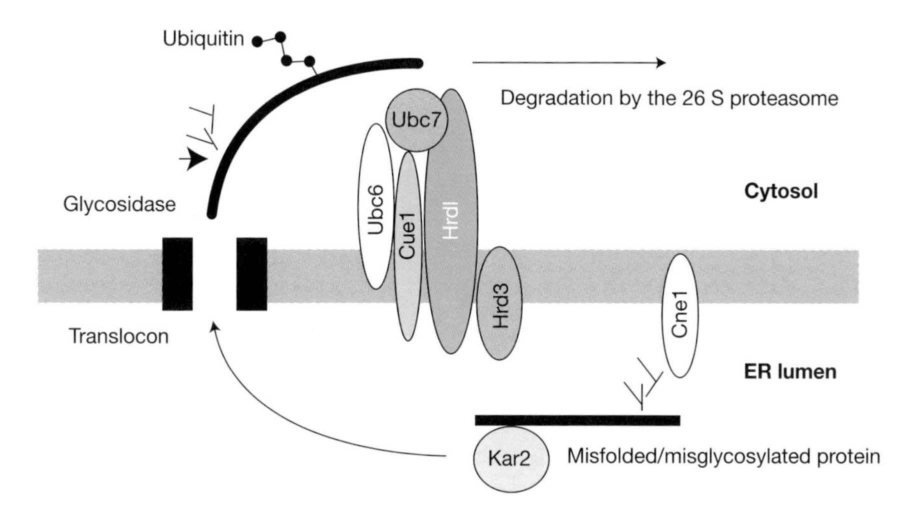


**Figure 6. ER-associated degradation: ubiquitin–proteasome degradation of ER proteins in *S. cerevisiae***

Misfolded or misglycosylated proteins are recognized by Cne1 (calnexin in mammalian cells) or Kar2 (BiP, the ER-resident 70 kDa chaperone/heat-shock protein in mammals), which probably prevents aggregation and presents the protein to the translocon for reverse translocation (dislocation) to the cytoplasm. A cytosolic glycosidase activity appears to trim or remove carbohydrate moieties and the protein is ubiquitylated by a complex including a RING-finger E3/ligase (Hrd1) which recruits the E2 UBc7. Cue1 also seems to be involved in E2 recruitment. The membrane E2 (Ubc6) may also be involved in ubiquitylation in ERAD. Hrd3, which extends into the ER lumen, serves to regulate Hrd1 activity.

### Antigen processing

The immune surveillance system in vertebrate cells involves the presentation of small peptides (8–10 amino acids) at the cell surface bound to the MHC I. The peptide–MHC I complex can then be recognized by cytotoxic T-cells. MHC I binds peptides that are derived by proteolysis of both cellular and viral proteins. These peptides are generated from ubiquitylated proteins by the proteasome in the cytoplasm and are transported into the ER by TAP (transporter associated with antigen processing) where they bind to MHC I and are trafficked to the cell surface. The proteolytic generation of peptides for presentation requires a special proteasome complex, the immunoproteasome, which generates peptides of the right length with the basic or hydrophobic C-terminal residues required for binding to MHC I. In the immunoproteasome 20 S core, catalytic subunits β1, β2 and β5 are replaced by interferon-inducible subunits β1i, β2i and β5i. In addition, the 20 S proteasome core is capped at one end by a 19 S regulator complex, but at the other by the proteasome activator PA28 (11 S) regulator complexes. The presence of the 11 S regulator and the interferon inducible 20 S subunits regulates the immunoproteasome in such a way that it generates the peptides required for antigen presentation [24]. Defective ribosomal products (see above) may provide a pool of self peptide antigens via the ubiquitin–proteasome pathway. Some viruses code for proteins that interact with the proteasome, but the effect of these interactions on antigen processing is not clear.

## Ubiquitin pathway and disease

### Central regulatory pathway involved in disease

It should be apparent by now that the ubiquitin pathway of intracellular proteolysis, not to mention the other functions of ubiquitin and ubiquitin paralogues, add a new layer on the cake of cellular physiology. Therefore, as already indicated, abnormalities in the pathway can lead to diseases like cancer, and pathogens such as viruses can exploit the pathway in many ways. Mutations in a HECT-domain E3 result in Angelman's syndrome [25], a neurological developmental disorder, and mutations in the RING-finger E3 *Parkin* cause juvenile onset recessive parkinsonism [26]. In addition, neurodegenerative diseases, e.g. Alzheimer's disease, Parkinson's disease, motor neuron disease and the spongiform encepalopathies, are characterized by the accumulation of aggregates of insoluble proteins [27]. These protein deposits contain ubiquitylated proteins and may be the consequence of failed attempts by the cell to dispose of abnormal proteins [28]. Indeed, in the aging brain, post-transcriptional 'molecular misreading' leads to increased quantities of a mutant C-terminal-extended ubiquitin that can still be incorporated into ubiquitin chains and is a potent inhibitor of the 26 S proteasome [29]. This impairment of the ubiquitin system may help to explain why these diseases are age-related. Once protein aggregates start to form, they can result in further inhibition of the proteasome [30], leading to accelerated aggregate formation as abnormal proteins fail to be degraded.

## Conclusion

Protein degradation is flexible, adaptable and participates in diverse cellular regulatory mechanisms. Undoubtedly, the major role for ubiquitin itself is in targeting proteins for degradation. However, recent research has shown that ubiquitylation targets proteins for degradation by both major cellular proteolytic systems, the 26 S proteasome and the lysosomal apparatus. Non-degradative roles for ubiquitin in protein regulation and vesicular trafficking are also emerging. The roles of the ubiquitin-like proteins SUMO and Nedd8 are being elucidated, but the roles of the various other ubiquitin-like proteins still remain to be determined. Every week sees the identification of new substrates for ubiquitin–proteasome-mediated degradation, and the ever-expanding repertoire for this pathway in the control of cellular processes by regulated proteolysis looks set to continue for some time. It is likely that every cellular pathway feels the bite of the proteasome.

### Summary

- *Ubiquitylation is a reversible modification that targets proteins for a variety of fates, including degradation by the 26 S proteasome.*
- *The 26 S proteasome is a large multi-subunit threonine protease, consisting of a 20 S catalytic core and 19 S regulator complexes, that degrades ubiquitylated proteins*
- *Many cell surface proteins are ubiquitylated and targeted to the lysosome for degradation. Mono-ubiquitylation can also serve as a signal for endocytosis.*
- *Signals in target proteins are recognized by ubiquitin ligase enzymes. Many ligases are complexes, with substrate recognition and ubiquitin transfer activities provided by different subunits. Phosphorylation precedes ubiquitylation in many cases.*
- *A system with similarities to the ubiquitylation pathway is responsible for controlling autophagic sequestration of cytoplasmic proteins for lysosomal degradation.*

We thank the Wellcome Trust, the Biotechnology and Biological Sciences Research Council, EU Framework IV, the Royal Society and the Alzheimer's Research Trust for support of work which led to some of the findings and concepts described in this essay.

## References


1. Vijay-Kumar, S., Bugg, C.E. & Cook, W.J. (1987) Structure of ubiquitin refined at 1.8 A resolution. *J. Mol. Biol.* **194**, 531–544
2. Yeh, E.T.H., Gong, L.M. & Kamitani, T. (2000) Ubiquitin-like proteins: new wines in new bottles. *Gene* **248**, 1–14
3. Hershko, A., Ciechanover, A. & Varshavsky, A. (2000) The ubiquitin system. *Nat. Med.* **6**, 1073–1081
4. Amerik, A.Y., Li, S.J. & Hochstrasser, M. (2000) Analysis of the deubiquitinating enzymes of the yeast Saccharomyces cerevisiae. *Biol. Chem.* **381**, 981–992
5. Voges, D., Zwickl, P. & Baumeister, W. (1999) The 26S proteasome: a molecular machine designed for controlled proteolysis. *Annu. Rev. Biochem.* **68**, 1015–1068
6. Fenteany, G., Standaert, R.F., Lane, W.S., Choi, S., Corey, E.J. & Schreiber, S.L. (1995) Inhibition of proteasome activities and subunit-specific amino-terminal threonine modification by lactacystin. *Science* **268**, 726–731
7. Seemuller, E., Lupas, A., Stock, D., Lowe, J. Huber, R. & Baumeister, W. (1995) Proteasome from Thermoplasma acidophilum — a threonine protease. *Science* **268**, 579–582
8. Groll, M., Ditzel, L., Lowe, J., Stock, D., Bochtler, M., Bartunik, H.D. & Huber, R. (1997) Structure of 20S proteasome from yeast at 2.4 angstrom resolution. *Nature (London)* **386**, 463–471
9. Hartmann-Petersen, R., Tanaka, K. & Hendil, K.B. (2001) Quaternary structure of the ATPase complex of human 26S proteasomes determined by chemical cross-linking. *Arch. Biochem. Biophys.* **386**, 89–94
10. Hofmann, K. & Falquet, L. (2001) A ubiquitin-interacting motif conserved in components of the proteasomal and lysosomal protein degradation systems. *Trends Biochem. Sci.* **26**, 347–350
11. Coffino, P. (2001) Antizyme, a mediator of ubiquitin-independent proteasomal degradation. *Biochimie* **83**, 319–323
12. Touitou, R., Richardson, J., Bose, S., Nakanishi, M., Rivett, J. & Allday, M.J. (2001) A degradation signal located in the C-terminus of p21(WAF1/CIP1) is a binding site for the C8 alpha-subunit of the 20S proteasome. *EMBO J.* **20**, 2367–2375
13. Rechsteiner, M., Realini, C. & Ustrell, V. (2000) The proteasome activator 11 S REG (PA28) and class I antigen presentation. *Biochem. J.* **345**, 1–15
14. Higashitsuji, H., Itoh, K., Nagao, T., Dawson, S., Nonoguchi, K., Kido, T., Mayer, R.J., Arii, S. & Fujita, J. (2000) Reduced stability of retinoblastoma protein by gankyrin, an oncogenic ankyrin-repeat protein overexpressed in hepatomas. *Nat. Med.* **6**, 96–99
15. Berezutskaya, E. & Bagchi, S. (1997) The human papillomavirus E7 oncoprotein functionally interacts with the S4 subunit of the 26 S proteasome. *J. Biol. Chem.* **272**, 30135–30140
16. Tyers, M. & Jorgensen, P. (2000) Proteolysis and the cell cycle: with this RING I do thee destroy. *Curr. Opin. Genet. Dev.* **10**, 54–64
17. Wang, C., Deng, L., Hong, M., Akkaraju, G.R., Inoue, J. & Chen, Z.J.J. (2001) TAK1 is a ubiquitin-dependent kinase of MKK and IKK. *Nature (London)* **412**, 346–351
18. Ciechanover, A., Gonen, H., Bercovich, B., Cohen, S., Fajerman, I., Israel, A., Mercurio, F., Kahana, C., Schwartz, A.L., Iwai, K. & Orian, A. (2001) Mechanisms of ubiquitin-mediated, limited processing of the NF-kappa B1 precursor protein p105. *Biochimie* **83**, 341–349
19. Hoppe, T., Matuschewski, K., Rape, M., Schlenker, S., Ulrich, H.D. & Jentsch, S. (2000) Activation of a membrane-bound transcription factor by regulated ubiquitin/proteasome-dependent processing. *Cell* **102**, 577–586
20. Meacham, G.C., Patterson, C., Zhang, W.Y., Younger, J.M. & Cyr, D.M. (2001) The Hsc70 co-chaperone CHIP targets immature CFTR for proteasomal degradation. *Nat. Cell Biol.* **3**, 100–105
21. Ward, C.L., Omura, S. & Kopito, R.R. (1995) Degradation of Cftr by the ubiquitin-proteasome pathway. *Cell* **83**, 121–127
22. Ng, D.T.W., Spear, E.D. & Walter, P. (2000) The unfolded protein response regulates multiple aspects of secretory and membrane protein biogenesis and endoplasmic reticulum quality control. *J. Cell Biol.* **150**, 77–88

23. Piguet, V., Schwartz, O., Le Gall, S. & Trono, D. (1999) The downregulation of CD4 and MHC-I by primate lentiviruses: a paradigm for the modulation of cell surface receptors. *Immunol. Rev.* **168**, 51–63
24. Kloetzel, P.-M. (2001) Antigen processing by the proteasome. *Nat. Rev. Mol. Cell Biol.* **2**, 179–187
25. Kishino, T., Lalande, M. & Wagstaff, J. (1997) UBE3A/E6-AP mutations cause Angelman syndrome. *Nat. Genet.* **15**, 70–73
26. Kitada, T., Asakawa, S., Hattori, N., Matsumine, H., Yamamura, Y., Minoshima, S., Yokochi, M., Mizuno, Y. & Shimizu, N. (1998) Mutations in the parkin gene cause autosomal recessive juvenile parkinsonism. *Nature (London)* **392**, 605–608
27. Lowe, J., Lennox, G., Mayer, R.J., Morrell, K., Landon, M. & Godwinausten, R.B. (1988) Ubiquitin immunoreactivity in neurodegenerative disease. *J. Neurol. Neurosurg. Psychiatry* **51**, 1360
28. Layfield, R., Alban, A., Mayer, R.J. & Lowe, J. (2001) The ubiquitin protein catabolic disorders. *Neuropathol. Appl. Neurobiol.* **27**, 171–179
29. Lam, Y.A., Pickart, C.M., Alban, A., Landon, M., Jamieson, C., Ramage, R., Mayer, R.J. & Layfield, R. (2000) Inhibition of the ubiquitin–proteasome system in Alzheimer's disease. *Proc. Natl. Acad. Sci. U.S.A.* **97**, 9902–9906
30. Bence, N.F., Sampat, R.M. & Kopito, R.R. (2001) Impairment of the ubiquitin–proteasome system by protein aggregation. *Science* **292**, 1552–1555

# 6

# Methionine aminopeptidases and angiogenesis

**Ralph A. Bradshaw[1] and Elizabeth Yi**

*Department of Physiology and Biophysics, College of Medicine, University of California, Irvine, CA 92697, U.S.A.*

## Abstract

The initiator methionine residue of proteins is removed during synthesis by a specific and ubiquitous enzyme, methionine aminopeptidase (MetAP). Prokaryotes have a single gene, while eukaryotes have two isoforms. This family of metalloenzymes generally cleaves substrates in which the penultimate residue is one of the seven smaller amino acids (glycine, alanine, serine, threonine, proline, cysteine and valine). One of the eukaryotic isoforms (MetAP2) has an additional non-proteolytic function and is the principle target of a family of anti-angiogenic drugs that are related to fumagillin. The resulting covalent modification inhibits the protease activity of MetAP2 and blocks cell-cycle function in endothelial and some cancer cells. The role of MetAP2 in the mitogenic activity of these cells is unknown.

## Introduction

It has long been appreciated that protein function is heavily dependent on co- and post-translational modifications that modulate or regulate cellular translocation, activity and degradation [1]. Indeed, such modifications, which are likely to include more than 200 distinct chemical reactions when they are all classified, generally must be identified directly, although at least some sites can be predicted with some accuracy from sequence data alone. The earliest modifications to a protein, in terms of its lifetime, are those that occur at or

[1]*To whom correspondence should be addressed (e-mail: rablab@uci.edu).*

near the N-terminus [2]. With the exception of side-chain modifications, in eukaryotes these involve limited proteolysis to remove one or more amino acids and/or modification of the α-amino group. For the most part, these reactions are thought to occur during protein synthesis, but there are certainly N-terminal modifications that happen after a protein is fully translated. The proteolytic cleavages are either exoprotease- or endoprotease-catalysed and generally involve the removal of the initiator methionine or a signal peptide respectively. The latter are primarily encountered in proteins that are transported across membranes, and specific enzymes found in the targeted compartments are responsible for their removal. There are also 'activation' peptides that are removed from the N-terminus, particularly from hormone and protease precursors, but these are usually later events and often accompany the export of the protein from the cell.

The removal of the initiator methionine is catalysed by a quite selective family of methionine aminopeptidases (MetAPs) that apparently occur ubiquitously. Methionine removal is considered to be largely a co-translational event in eukaryotes, generally occurring when nascent chains are approx. 20–40 residues in length; however, this may not always be the case. In prokaryotes, there is an obligatory deformylation step before the cleavage of the methionine can occur because protein synthesis in these organisms is initiated with *N*-formylmethionine. Both reactions take place early in the lifetime of the protein but it is not entirely clear whether they are co- or post-translational (or both). The general substrate profile that governs initiator methionine removal is universal and depends largely on the adjacent amino acid [2]. The modification of α-amino groups is more diverse and there are over a dozen such modifications that are known [1]. By far the most common of these is $N^{\alpha}$-acetylation, which is catalysed by $N^{\alpha}$-acetyltransferases [2].

The combined effect of the MetAPs and $N^{\alpha}$-acetyltransferases in eukaryotes is to produce four classes of N-termini among cytosolic/nuclear proteins. The methionine residue is generally removed from proteins where the adjacent amino acid is small (glycine, alanine, serine, threonine, cysteine, proline and valine) and four of these (glycine, alanine, serine and threonine) are then acetylated to a substantial degree. Of the larger group that retains the methionine residue, those proteins with aspartate, glutamate or asparagine residues in the adjacent position also are usually acetylated on the α-amino group of methionine (Figure 1) [2]. Three of the residues (valine, asparagine and glutamine) are in a 'grey zone', particularly in *in vitro* experiments, and may or may not be modified. In this article, we review the specific MetAP family involved in processing initiator methionine residues, with special attention being paid to the eukaryotic proteins, which occur in two major isoforms [3]. Although they share a common ancestor and have similar catalytic properties, the isoforms clearly enjoy some unique physiological activities. In endothelial cells, one of them is required for cell cycle passage and has thus become an important target for inhibiting angiogenesis in oncologic therapy.

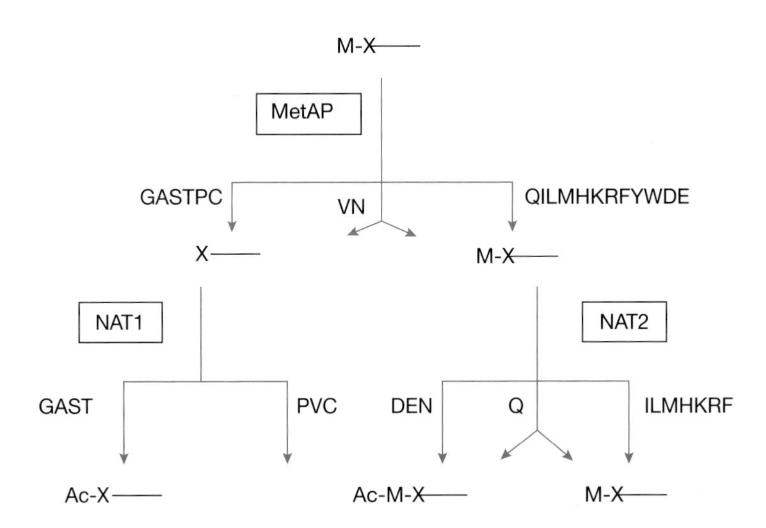


**Figure 1. General pathway for the co-translational N-terminal processing of eukaryotic proteins by MetAP and N$^{\alpha}$-acetyltransferase (NAT)**
The scheme does not distinguish separate roles for MetAPI and MetAP2. NATI and NAT2 are arbitrary designations for the two main classes of known NATs. M, initiator methionine; X, second amino acid residue; Ac, acetyl.

## MetAPs: physical and chemical properties

Aminopeptidases are widely encountered in Nature and are involved in a number of biological processes such as maturation, regulation and degradation [4]. Generally, they are metalloenzymes (usually zinc) and can have monomeric or multimeric quaternary structures. In terms of specificity, they are usually grouped in two basic classes: those that remove a range of large hydrophobic residues from substrates in a processive manner, and those that remove specific residue types. Leucine aminopeptidase is an example of the former class, while MetAP is a member of the latter. The general structure and mechanism of each type is well known. They do not share any sequence or structural similarity, but there are some interesting similarities at the mechanistic level. It may be presumed that they exemplify convergent evolution. During recent years, much about the structure, function and evolution of the MetAPs has been determined [5,6] and is summarized briefly below.

### Isolation

Although it has been appreciated for over 30 years that protein synthesis in eukaryotes is initiated by methionine, it was not immediately understood that this residue is removed in most mature proteins. Although there were several interim reports describing aminopeptidase activity against methionine substrates (summarized in [7]), it was not until the early 1980s that a profile of initiator methionine removal/retention, based on the nature of the penultimate residue, began to emerge from database analyses and the more systematic examination of mutant iso-cytochrome *c* molecules in yeast [8]. These analyses correctly presaged the general specificity of the enzymes really involved, i.e. the penultimate residue was one of the seven smallest amino acids (Figure 1), which was subsequently confirmed by direct measurements (summarized in [2]). This information was essential to the isolation and characterization of MetAPs, first from *Escherichia coli* [9] and then from yeast [10]. The prokaryotic form was found to be a monomeric protein of approx. 30 kDa, whose activity was strongly stimulated by $Co^{2+}$. The yeast enzyme had the same properties, including quite significant sequence identity to the prokaryotic catalytic domain. It also had an N-terminal extension of approx. 12 kDa, containing two putative zinc fingers. However, when the enzyme was isolated from porcine liver [11], it was markedly different in that it had an apparent molecular mass of 67 kDa. Furthermore, when the sequence of the corresponding human enzyme was determined [3], it did not show strong sequence identity with either the yeast or bacterial moieties and lacked any evidence of the zinc fingers. It did exhibit the same substrate pattern and was also dependent on $Co^{2+}$. The structure of the *E. coli* enzyme [12], which revealed a novel 'pitta bread' fold that suggested an internal homology arising from an early gene duplication event, provided the insight to resolve this conundrum. When the ligands for the two metal ions bound in the putative catalytic site, identified in the structural studies, were aligned with the human sequence, there was a complete match (albeit it required a substantial insertion to align the last ligand), suggesting that a structural relationship did actually exist. Additional database and sequence analyses, as well as model building, led to the correct prediction that there were two classes of MetAPs in eukaryotes (now denoted 1 and 2) [3]. The MetAP1 family is related to the *E. coli* enzyme [9] and the MetAP2 group is related to the enzyme first isolated from porcine liver [11]. Interestingly, eubacteria contain only the catalytic portion of the type 1 protein, while archaebacteria contain only a similar truncated version of the type 2 enzyme.

## Structure

The overall organization of the MetAP family is shown schematically in Figure 2. The principal structural features distinguishing the two subforms are the N-terminal domains and the catalytic domain insert in the type 2 family. The former sequences of both isoforms are quite different in yeast and human (Figure 3) and their functional significance, or why they are absent from the prokaryotic enzymes, is unknown. It has been suggested that the zinc fingers in the type 1 protein may tether the enzyme to the surface of the ribosome, where it can bind and process nascent polypeptide chains as they are synthesized [5]. Both yeast type 1 and human type 2 MetAPs are catalytically active without this domain and the truncated yeast protein will function *in vivo*, albeit less effectively than the native version. The N-terminal sequence of the type 2 proteins is unusually highly charged, containing both poly(acidic) and poly(basic) sequences. The human domain is considerably longer (154 compared with 86 residues) than that of yeast, which lacks a highly negatively charged internal sequence, and this may be important for unique functions of MetAP2 characteristic of higher eukaryotic cells. This domain was not visible in the X-ray structural analysis of human MetAP2 and, therefore, its conformation is unknown. It too may act as a tether to the ribosome or to an organelle, or it may have an entirely different function. The structure of the catalytic domain insert is known for both the human and *Pyrococcus furiosus* (an archaebacterium) enzymes ([6] and references cited therein) and it is composed of a largely helical loop located on the surface of the C-terminal portion of the molecule. Its impact on MetAP function also remains to be determined. A quite plausible scheme to account for the evolutionary development of these structures and their distribution in organisms has been proposed [13]. As shown in Figure 4, the earliest form of the enzyme is likely to have appeared after a gene duplication event and it served as a precursor to both present-day subtypes. Of course, one or the other of the prokaryotic forms may have given rise to the other by the insertion or deletion of the catalytic domain insert. The expression of both MetAPs in modern eukaryotes ultimately resulted from the endosymbiotic development of the mitochondrion, an event that would have been succeeded by the presumably independent acquisition of the N-terminal domains by each gene.

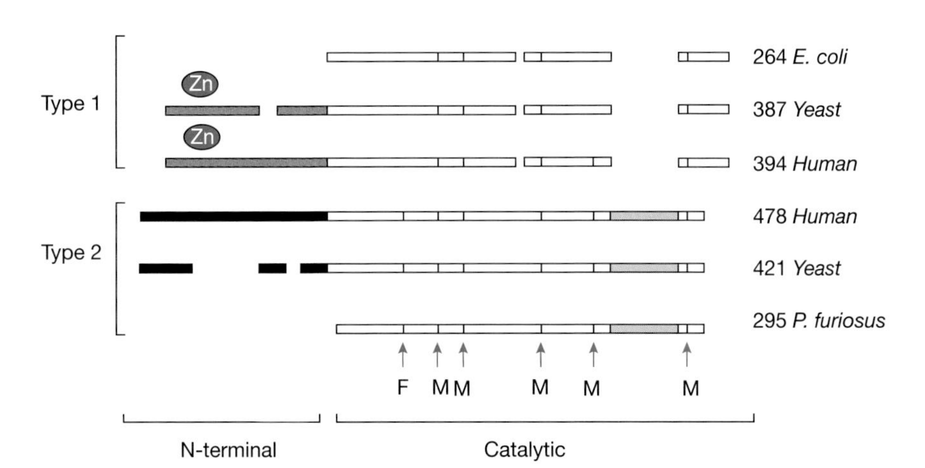


**Figure 2. Schematic presentation of the structural organization of the MetAP family**
Individual domains are indicated by various shadings and are presented approximately to scale. Black, N-terminal extension characteristic of MetAP2s; grey, N-terminal extension characteristic of MetAP1s; blue, catalytic domain insert characteristic of MetAP2s. Major deletions (and insertions) deduced from sequence and structural comparisons are indicated as gaps. M, site of a metal ligand; F, location of the histidine residue modified in type 2 enzymes; Zn, zinc-ion-binding site (zinc fingers). The number of residues for each protein is given in the column at the right.

```
(a)

   MALFQRAGSMAAVETRV-CETDGCSSEA--KLQCPTCIKLGIQGSYFCSQEC
MSTATTTVTTSDQASHPTKIYCSGLQCGRETSSQMKCPVCLKQGI-VSIFCDTSC

FKGSWATHKLLHKKAKDEKAKREVSSWTVEGDINTDPWAGYRYTGKLRPHYPLMP
YENNYKAHKALHN-AKDG----------LEGAY--DPFPKFKYSGKVKASYPLTP

TRPVPSYIQRPDYADHPLGMSESEQALKGT    (H. sapiens I)     (133aa)
RRYVPEDIPKPDWAANGLPVS            (S. cerevisiae I)   (116aa)


(b)

MAGVEEVAASGSHLNG-DLDPDDREEGAASTAEEAAKKKRRKKKKSKGPSAAGEQEPDKE
MTDAEIENSPASDLKELNLENEGVEQQDQAKADESDPVESKKKKN---------------

SGASVDEVARQLERSALEDKERDEDDEDGDGDGDGATGKKKKKKKKKRGPKVQTDPPSVP
---------------------------------------KKKKKKKSNVKKIEL------

ICDLYPNGVFPKGQECEYPPTQDGRTAAWRTTSEEK   (H. sapiens II)   (154aa)
---LFPDGKYPEGAWMDYHQDFNLQ----RTTDE     (S. cerevisiae II) (86aa)
```

**Figure 3. N-terminal domain sequences of MetAP1 and MetAP2**
(a) N-terminal zinc-finger domain sequences from human and yeast MetAP1. (b) N-terminal poly(acidic)/(basic) domain sequences from human and yeast MetAP2.

## Metal usage

All of the MetAP isoforms require metal ions to function, but there is some uncertainty as to the number and type [5] in various species. Originally, the bacterial form was found to be maximally stimulated by high concentrations of $Co^{2+}$ [9] and this was confirmed with the yeast and porcine liver enzymes [10,11], leading to the general conclusion that these were $Co^{2+}$-dependent enzymes. Two cobalt atoms were bound at the putative catalytic site [12], suggesting that they were members of the larger binuclear metallohydrolases. However, subsequent studies with the recombinant enzyme have challenged this view. Walker and Bradshaw [14] demonstrated that yeast MetAP1 could utilize $Zn^{2+}$ equally well at low concentrations and that, in the presence of physiologically relevant amounts of glutathione, the $Co^{2+}$-containing enzyme lost activity. D'souza and Holz [15] found that recombinant *E. coli* MetAP (also a type 1 enzyme) was equally active with $Fe^{2+}$ or $Co^{2+}$ and suggested that, under physiological conditions (strictly anaerobic), the former would likely to be the metal ion utilized. The type 2 isotypes from archaebacteria and humans seem to function strictly with $Co^{2+}$, although the latter has not been tested under anaerobic conditions or with physiological amounts of thiol present [5]. The apparent metal usage of the various MetAPs is summarized in Table 1. The low abundance of $Co^{2+}$ and the scarcity of $Co^{2+}$-dependent enzymes in eukaryotes (other than cobalamin-dependent enzymes, there are no other known $Co^{2+}$-containing enzymes besides the MetAPs [16]) argue against this metal ion and in favour of zinc or possibly something else. It has been suggested that this enzyme family may be unusual in that different

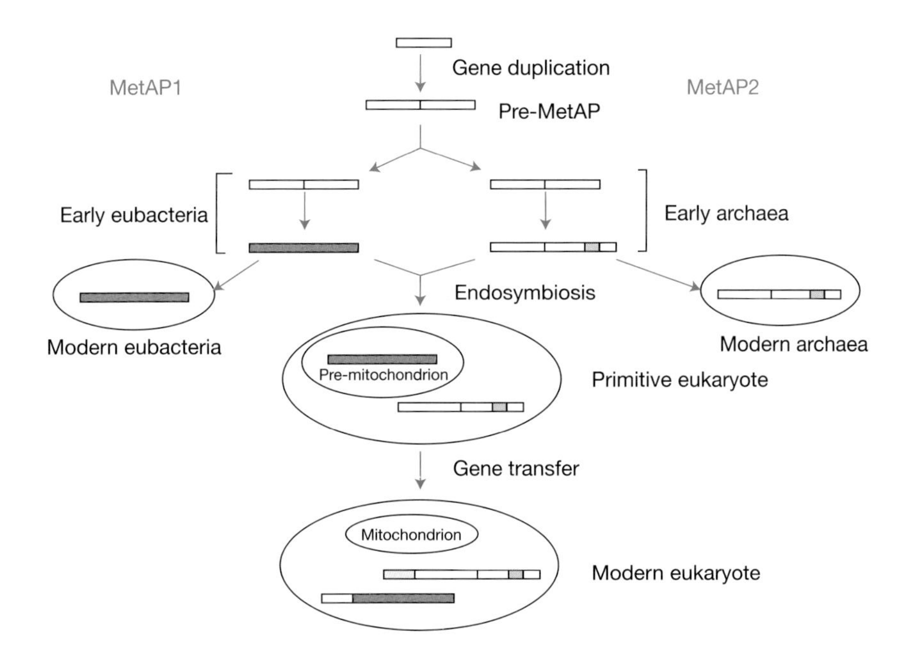


**Figure 4. Proposed path of evolution of the MetAP family**
Shading indicates distinguishing domains of each protein acquired during the definitive stages of the process.

members in different species and kingdoms may use different metal ions [6]. In fact, it is possible that the ion used may actually vary in any given species depending on growth and other environmental conditions.

## Substrate specificity

Although the general specificity of all the MetAPs is well preserved (Figure 1), there is accumulating evidence that there are important nuances that can affect individual substrates, particularly with respect to subtype utilization. In one study, Walker and Bradshaw [17] demonstrated that recombinant yeast MetAP1 cleaved the peptide MXSHRWDW as well with asparagine as the second residue as it did with valine, but that neither was efficiently turned over. They suggested that proteins with these dipeptide sequences at the N-terminus would probably be found both with and without methionine regardless of additional influences. In the same study, these workers also showed that substitution of two $S_1'$ pocket residues ($Met^{329}$ and $Gln^{356}$) with alanine resulted in derivatives that could efficiently cleave a broader range of substrates. Only the charged residues (aspartate, glutamate, lysine and arginine) and tyrosine in the adjacent position remained recalcitrant. Turk et al. [18] made a direct comparison of the human isotypes and observed clear differences both *in situ* (taking advantage of specific inhibition of MetAP2) and in *in vitro* studies with model peptides. However, as shown in the latter

**Table 1. Metal co-factor utilization by MetAPs**

RSH, reducing agent, e.g. mercaptan.

| MetAP | −RSH | +RSH |
|---|---|---|
| *E. coli* | $Co^{2+}$, $Fe^{2+}$ | $Co^{2+}$, $Zn^{2+}$, $Fe^{2+}$ |
| Yeast (type I) | $Co^{2+}$, $Zn^{2+}$ | $Zn^{2+}$ |
| *P. furiosus* | $Co^{2+}$ | $Co^{2+}$ |
| Human (type 2) | $Co^{2+}$ | ? |

studies, with one exception the differences were only in the kinetic constants. Moreover, the conclusions regarding preferred substrates for each human isotype were in marked contrast to other observations for the yeast enzymes [10,17]. Although these differences may reflect species variations to some degree, it is likely that they also reflect the influence of downstream residues. The one exception was the N-terminal sequence of glyceraldehyde-3-phosphate dehydrogenase (MVKVGVNG). Human MetAP1 did not hydrolyse this peptide but human MetAP2 did, consistent with the observation that the processing of this protein in cells in which MetAP2 had been specifically inhibited was impaired [18]. However, the N-terminal peptide of haemoglobin (MVHTLPEE) was cleaved by both enzymes and, when normalized to MGMM, isotype 1 was actually more efficient than isotype 2. These observations suggest that there are additional factors that influence the general specificity of the two isotypes of MetAP in eukaryotic cells and that these differences are likely to be important in defining their exact physiological roles.

## Physiological properties

The principal function of the MetAPs is to remove initiator methionine, as it is almost certainly a requirement of all living cells to keep the pool of free methionine from being exhausted from this one activity alone. Methionine plays many roles in cells beside the initiation of protein synthesis, and the universal existence of these enzymes suggests that this activity developed very early in evolution. It is known from studying yeast null mutations that the two yeast enzymes are sufficiently redundant that it can survive with either one alone, but that the double null mutation is lethal. The inhibition of MetAP2 in endothelial cells causes cell-cycle arrest, but is otherwise tolerated. Bacteria require the single form present for survival. It is more difficult, however, to explain why the substrate specificity was limited to the smaller amino acids (and has remained unchanged) or why eukaryotes retained both isotypes after the incorporation of the prokaryotic mitochondrial precursor. While the answers to these intriguing questions are still unclear, a brief review of the physiological activities associated with the MetAPs, beyond simply methionine recovery, provides perhaps some insight.

**Protein turnover**

In eukaryotes, the substrate profile of the MetAPs matches quite well with the profile of destabilizing residues identified as directing intracellular N-terminal-mediated proteolysis (the N-end Rule) [19]. This was initially determined in yeast using a fused gene construct composed of ubiquitin and an altered β-galactosidase. When the residues in the penultimate position that normally were not processed by MetAP were exposed via the cleavage of ubiquitin, they were recognized by an N-end-Rule-specific E3 ligase, polyubiquitinated and rapidly destroyed by the proteasome (see Chapter 5). In contrast, residues that were normally exposed by MetAPs (the smaller amino acids) were quite stable to N-end-Rule-mediated degradation. Thus, one reason that the MetAP specificity has been retained is to prevent premature degradation by the N-end Rule apparatus. However, clearly the N-end Rule has evolved to accommodate the MetAP substrate specificity and not vice versa because the N-end Rule, such as it is in prokaryotes, is different than in eukaryotes [19], whereas the MetAP specificity is not.

The existence of a quasi-stable class of proteins resulting from the combination of these two processing pathways suggested the possibility of a potentially substantial subset of proteins whose turnover could be controlled by the subsequent or 'downstream' removal of the protective N-terminal methionine. This, however, does not appear to be the case. There are only a few examples of N-end-Rule-mediated degradation of an identified protein and these resulted from internal cleavages; there are no examples of protein turnover by the N-end Rule initiated by exopeptidase (or acylamino acid hydrolase) activity. Therefore, the connection between MetAP activity and the N-end Rule appears to be a prophylactic one rather than a means to programmed protein turnover.

**Regulation of translation**

Prior to the isolation of porcine MetAP2, the isolation and characterization of a protein that inhibited the phosphorylation of eukaryotic initiation factor 2α (eIF2α) was reported ([7] and references cited therein). The effect of this inhibition is to promote translation. The protein had an apparent molecular mass of 67 kDa (and was thus designated p67) and was reported to contain 12 O-linked monosaccharides. However, its calculated mass (without the sugar residues) was approx. 53 kDa. A comparison of several peptides from porcine MetAP2 with the sequence of rat p67 clearly indicated that they were the same protein. Subsequent sequence analyses of the human protein [3] confirmed this and did indicate a likely sequencing error that affected several residue assignments in the C-terminal region of the rat protein.

The relationship of this second activity to the catalytic function is unclear. Specific inhibition of the MetAP2 activity (see below) does not affect the activity of eIF2α, which presumably is related to its ability to interact with one or more components of the eIF complex. Furthermore, this inhibitory activity apparently requires the glycosylation. Deglycosylated p67 is not active nor does it bind to a lectin column [7]. Interestingly, none of the recombinant human MetAP2 preparations have been reported to be modified in this fashion. However, the human N-terminal sequence contains over a dozen potential sites for this modification, including the long segment that is not present in yeast (Figure 3), and this would provide a possible role for the N-terminal domain, at least in higher eukaryotes.

### Inhibition by fumagillin-like compounds

The finding that MetAP2 was the principal target for a previously identified class of drugs with anti-amoebic and immunosuppressive activity was certainly unexpected. It had also been established that that these compounds inhibited endothelial cell division and were thus anti-angiogenic also [5]. Two groups [20,21] independently showed that fumagillin and ovalicin (as well as other derivatives) (Figure 5) covalently modified MetAP2, resulting in inactivation of the catalytic function but not the inhibition of phosphorylation. The reactive moiety was shown to be the ring epoxide and the site of modification was found to be the imidazole side chain of $His^{231}$ (in human MetAP2) (see Figure 2). It was shown subsequently that the *E. coli* enzyme (a type 1 isoform) could also be modified when much higher concentrations of reagent were used; the site of modification was $His^{79}$, which occupies a similar position to $His^{231}$ in the human protein [6]. The differences in reactivity relate to the differences in sequence and main chain folding that characterize the two isoforms.

These observations are important for several reasons. First, the specific inhibition of MetAP2 in human cells did not result in cell death-indicating that they can survive with only the type 1 enzyme; in this regard, they are similar to yeast and *Drosophila*. However, it remains unclear whether they would survive if MetAP1 was specifically inhibited instead. Secondly, the effects on cell-cycle progression (see also below) establish that the isoforms are not simply redundant with respect to their processing activity. Finally, they suggest that there is a specific set of substrates that is processed by MetAP2 (either wholly or in part), that leads to unique physiological responses. The particular sensitivity of endothelial cells to MetAP2 inhibition translates in that case to anti-angiogenesis.

## Angiogenesis

Angiogenesis, the process of developing new blood vessels from existing ones, is essential in the developing organism and continues in the adult in specialized tissues. Pathologically, it is also essential in such conditions as wound healing, arthritis, inflammation and tumour progression [22]. Folkman [23] pioneered the oncologic aspects of angiogenesis, establishing that this new blood supply was required for tumour expansion beyond a distance of a few millimetres. Angiogenesis is a highly regulated process that involves many substances and conditions, and can be induced by several growth factors such as fibroblast growth factor and vascular endothelial growth factor. The fortuitous

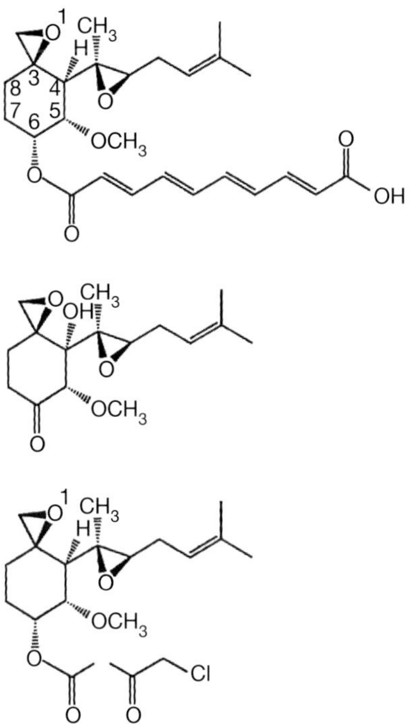

Fumagillin

Ovalicin

TNP-470

**Figure 5. Fumagillin and related compounds that act as anti-angiogenic drugs through the specific inhibition of endothelial cell MetAP2**

identification of the effect of fumagillin on endothelial cell division that resulted from a fungal contamination led to the development of a compound, AGM-1470 (also known as TNP-470), that was more potent than either fumagillin or ovalicin as an anti-angiogenic agent and showed reduced side effects [22]. It is being tested clinically at present as an anti-tumour drug. Recently, Han et al. [24] have reported a new cinnamic ester derivative that is a 1000 times more potent as an inhibitor of endothelial cell proliferation than TNP-470. It was actually the product of rational drug design utilizing the structure of human MetAP2, the known target.

Although the site of action of this group of drugs is now understood, it is not yet appreciated why the inhibition of MetAP2 causes cell-cycle arrest in endothelial cells. This is not due to any imbalance between the isoforms, as non-endothelial cells also expressed both MetAP1 and MetAP2 [25]. Thus, barring another as yet undiscovered activity like the inhibition of the phosphorylation of eIF2α, the role of MetAP2 in controlling cell-cycle progression in endothelial cells most probably resides in the subtle, but real, differences in substrate utilization already observed for the two isotypes [18]. Indeed, as summarized in [5], it was found that TNP-470 treatment caused cyclin D mRNA expression to be suppressed and inhibited the phosphorylation of retinoblastoma protein. The cyclin-dependent kinases, cdc2 and cdk2, which may be involved in retinoblastoma protein modification, were also indirectly inhibited. It has also been established that p53 and $p21^{WAF1/CIP1}$ are essential elements in the MetAP2-controlled process, as cells lacking these proteins were insensitive to the inhibitor. However, just which proteins are actually substrates for MetAP2 and why their putative lack of processing leads to these various responses is not presently known.

Endothelial cells are clearly quite sensitive to MetAP2 inhibition, as there is little or no effect of TNP-470 on the proliferation of astrocytes, dermal fibroblasts, epidermal keratinocytes, mammary and prostate epithelial cells, and umbilical artery muscle cells [25]. However, several cancer cell lines are inhibited directly by TNP-470, albeit at various concentrations of the drug, some of which were several orders of magnitude higher than that required for blocking endothelial cell growth [22].

## Conclusion

The findings described here, although still clearly incomplete, suggest a specialized role for MetAP2 relative to MetAP1 in controlling important cellular processes; however, there is certainly some redundancy. More detailed information on substrate specificity will provide better insight into this issue and may also allow the identification of key substrates by database mining. This in turn may allow the development of better drugs to control the pathologies that involve MetAP2 action.

### Summary

- *Initiator methionine removal is catalysed by a ubiquitous family of MetAPs, which comprises two isotypes (MetAP1 and MetAP2) that evolved from a common precursor. Prokaryotes have a single form; eukaryotes have extended versions of each.*
- *All MetAPs require metal cofactors, which may vary among species and kingdoms, and share the same general substrate profile.*
- *In addition to its exopeptidase activity, human (and other higher eukaryotic) MetAP2 also promotes translation by inhibiting the phosphorylation of eIF2$\alpha$.*
- *The catalytic, but not the inhibitory, activity of MetAP2 is blocked by fumagillin and related drugs; a single histidine is covalently modified. MetAP1 is modified only at much higher drug concentrations.*
- *The proliferation of endothelial cells and, to a lesser degree, several cancer cells is inhibited by these drugs and they are being tested clinically as potential anti-tumour (anti-angiogenic) compounds.*

- *Although several important cell cycle proteins are affected by MetAP2 inhibition, the actual unprocessed substrates that directly cause the anti-proliferative responses are unknown.*

The work described that emanated from the authors' laboratory was supported by a United States Public Health Service research grant DK 32465.

## References

1. Krishna, R.G. & Wold, F. (1998) Posttranslational modifications. In *Proteins: Analysis and Design* (Angeletti, R.H., ed.), pp. 121–206, Academic Press, Orlando, FL
2. Arfin, S.M. & Bradshaw, R.A. (1988) Cotranslational processing and protein turnover in eukaryotic cells. *Biochemistry* **27**, 7979–7984
3. Arfin, S.M., Kendall, R.L., Hall, L., Weaver, L.H., Stewart, A.E., Matthews, B.W. & Bradshaw, R.A. (1995) Eukaryotic methionyl aminopeptidases: two classes of cobalt-dependent enzymes. *Proc. Natl. Acad. Sci. U.S.A.* **92**, 7714–7718
4. Taylor, A. (ed.) (1996) *Aminopeptidases*, R.G. Landes Company, Austin, TX
5. Bradshaw, R.A., Hope, C.J., Yi, E. & Walker, K.W. (2001) Co- and postranslational processing: the removal of methionine. In *The Enzymes: Co- and Posttranslational Proteolysis of Proteins* (Dalbey, R.E. & Sigman, D.S., eds), vol. XXII, 3rd edn, pp. 389–420, Academic Press, San Diego, CA
6. Lowther, W.T. & Matthews, B.W. (2000) Structure and function of the methionine aminopeptidases. *Biochim. Biophys. Acta* **1477**, 157–167
7. Datta, B. (2000) MAPs and POEP of the roads from prokaryotic to eukaryotic kingdoms. *Biochimie* **82**, 95–107
8. Sherman, F., Stewart, J.W. & Tsunasawa, S. (1985) Methionine or not methionine at the beginning of a protein. *Bioessays* **3**, 27–31
9. Ben-Bassat, A., Bauer, K., Chang, S.Y., Myambo, K., Boosman, A. & Chang, S. (1987) Processing of the initiation methionine from proteins: properties of the Escherichia coli methionine aminopeptidase and its gene structure. *J. Bacteriol.* **169**, 751–757
10. Chang, Y.H., Teichert, U. & Smith, J.A. (1990) Purification and characterization of a methionine aminopeptidase from Saccharomyces cerevisiae. *J. Biol. Chem.* **265**, 19892–19897
11. Kendall, R.L. & Bradshaw, R.A. (1992) Isolation and characterization of the methionine aminopeptidase from porcine liver responsible for the co-translational processing of proteins. *J. Biol. Chem.* **267**, 20667–20673
12. Roderick, S.L. & Matthews, B.W. (1993) Structure of the cobalt-dependent methionine aminopeptidase from Escherichia coli: a new type of proteolytic enzyme. *Biochemistry* **32**, 3907–3912
13. Bradshaw, R.A., Brickey, W.W. & Walker, K.W. (1998) N-terminal processing: the methionine aminopeptidase and N alpha-acetyl transferase families. *Trends Biochem. Sci.* **23**, 263–267
14. Walker, K.W. & Bradshaw, R.A. (1998) Yeast methionine aminopeptidase I can utilize either $Zn^{2+}$ or $Co^{2+}$ as a cofactor: a case of mistaken identity? *Protein Sci.* **7**, 2684–2687
15. D'souza, V.M. & Holz, R.C. (1999) The methionyl aminopeptidase from Escherichia coli can function as an iron(II) enzyme. *Biochemistry* **38**, 11079–11085
16. Kobayashi, M. & Shimizu, S. (1999) Cobalt proteins. *Eur. J. Biochem.* **261**, 1–9
17. Walker, K.W. & Bradshaw, R.A. (1999) Yeast methionine aminopeptidase I. Alteration of substrate specificity by site-directed mutagenesis. *J. Biol. Chem.* **274**, 13403–13409
18. Turk, B.E., Griffith, E.C., Wolf, S., Biemann, K., Chang, Y.H. & Liu, J.O. (1999) Selective inhibition of amino-terminal methionine processing by TNP-470 and ovalicin in endothelial cells. *Chem. Biol.* **6**, 823–833
19. Varshavsky, A. (1997) The N-end rule pathway of protein degradation. *Genes Cells* **2**, 13–28

20. Griffith, E.C., Su, Z., Turk, B.E., Chen, S., Chang, Y.H., Wu, Z., Biemann, K. & Liu, J.O. (1997) Methionine aminopeptidase (type 2) is the common target for angiogenesis inhibitors AGM-1470 and ovalicin. *Chem. Biol.* **4**, 461–471
21. Sin, N., Meng, L., Wang, M.Q., Wen, J.J., Bornmann, W.G. & Crews, C.M. (1997) The antiangiogenic agent fumagillin covalently binds and inhibits the methionine aminopeptidase, Met AP-2. *Proc. Natl. Acad. Sci. U.S.A.* **94**, 2362–2367
22. Castronovo, V. & Belotti, D. (1996) TNP-470 (AGM-1470): mechanisms of action and early clinical development. *Eur. J. Cancer* **32A**, 2520–2527
23. Folkman, J. (1971) Tumor angiogenesis: therapeutic implications. *N. Engl. J. Med.* **285**, 1182–1186
24. Han, C., Ahn, S., Choi, N., Hong, R., Moon, S., Chun, H., Lee, S., Kim, J., Hong, C., Kim, D. et al. (2000) Design and synthesis of highly potent fumagillin analogues from homology modeling for a human MetAP-2. *Bioorg. Med. Chem. Lett.* **10**, 39–43
25. Wang, J., Lou, P. & Henkin, J. (2000) Selective inhibition of endothelial cell proliferation by fumagillin is not due to differential expression of methionine aminopeptidases. *J. Cell. Biochem.* **3**, 465–473

# 7

# Precursor convertases in the secretory pathway, cytosol and extracellular milieu


## Nabil G. Seidah[1] and Annik Prat

*Laboratory of Biochemical Neuroendocrinology, Clinical Research Institute of Montreal, 110 Pine Avenue West, Montreal, QC, Canada H2W 1R7*



### Abstract

Precursor proteins that transit through the secretory pathway often require processing at specific sites in order to release their bioactive entities. The most prevalent limited proteolysis occurs at single or paired basic residues, and is achieved by one or more of the seven subtilisin-like proprotein convertases (PCs); Furin, PC1, PC2, PACE4 (paired basic amino acid converting enzyme 4), PC4, PC5 and PC7. Other types of cleavages occur at hydophobic residues, some of which are performed by subtilisin/kexin-like isozyme-1 (SKI-1), which is also known as site-1 protease. Together, the PCs and SKI-1 regulate the activity of a large variety of cellular proteins, including growth factors, neuropeptides, receptors, enzymes and even toxins and glycoproteins from infectious retroviruses. These processing events are exquisitely regulated by multiple zymogen-activation steps, as well as by specific subcellular localization signals. The above mentioned convertases are implicated in a number of pathologies such as cancer, neurodegenerative diseases, endocrine disorders and inflammation. Recently, it was recognized that the metalloendopeptidase N-arginine dibasic convertase (NRDc; nardilysin), which cleaves at the N-terminus side of basic residues in dibasic pairs, is



[1]*To whom correspondence should be addressed (e-mail: seidahn@ircm.qc.ca).*

localized both in the cytosol and at the cell surface or in the extracellular milieu. While NRDc binds heparin-binding epidermal growth factor (HB-EGF) at the cell surface and potentiates its physiological effect, HB-EGF potently inhibits the NRDc's activity. NRDc could represent the equivalent of the PCs in the cytosol or the extracellular space.

## Introduction

While the number of functional genes within the human and rodent genomes has not yet been confirmed, it seems to be less than 100000. However, the limited repertoire of gene products is greatly diversified by post-translational modifications, such as those that modify a single amino acid, e.g. phosphorylation, amidation, sulphation, glycosylation, acetylation, carboxymethylation, oxidation or pyroglutaminyl formation. Others involve the generation of one or more cleavage products through regulated limited proteolysis of substrates in specific subcellular compartments. In the secretory pathway, such processing often occurs at specific single or paired basic residues. This chapter concentrates on the processing of polypeptides at basic sites by (i) a family of mammalian serine proteases and (ii) the metalloendopeptidase *N*-arginine dibasic convertase (NRDc; nardilysin), which is present in the cytosol and at the extracellular surface. In addition, we will deal with the properties of a novel subtilisin-like enzyme that does not cleave after basic residues, called subtilisin/kexin-like isozyme 1 (SKI-1) or site 1 protease (S1P), and define its critical importance in cellular homoeostasis and in viral infections.

## Proprotein convertases (PCs) and SKI-1, and their pathological implications

### Processing and activation of secretory precursors

Proproteins are the fundamental units from which bioactive proteins and peptides, in particular neuropeptides, are derived by limited proteolysis. Secretory precursors are usually cleaved at the general motif (Lys/Arg)-$(Xaa)_n$-(Lys/Arg)↓, where $n=0$, 2, 4 or 6 and Xaa is usually not Cys (↓ indicates the cleavage site). Seven mammalian PCs that cleave at the above motif have been identified since 1990 (Figure 1): furin, PC1 (also called PC3), PC2, PC4, PACE4 (paired basic amino acid converting enzyme 4), PC5 (also called PC6), and PC7 [also called PC8, LPC (lymphoma PC) or SPC7 (subtilisin-like PC)] [1–4]. These serine proteases belong to the yeast kexin subfamily of subtilases. Recently, a new member (SKI-1 or S1P), more closely related to pyrolysin than to kexin, was identified (Figure 1). It exhibits a specificity for cleavage at the motif (Arg/Lys)-Xaa-(hydrophobic)-(Leu/Thr/Lys/Phe)↓ (↓ indicates the cleavage site) [5]. These convertases are responsible for the tissue-specific processing of multiple precursors, which often involves a cascade of cleavage events that generates diverse bioactive molecules in an exquisitely regulated manner [1–4].

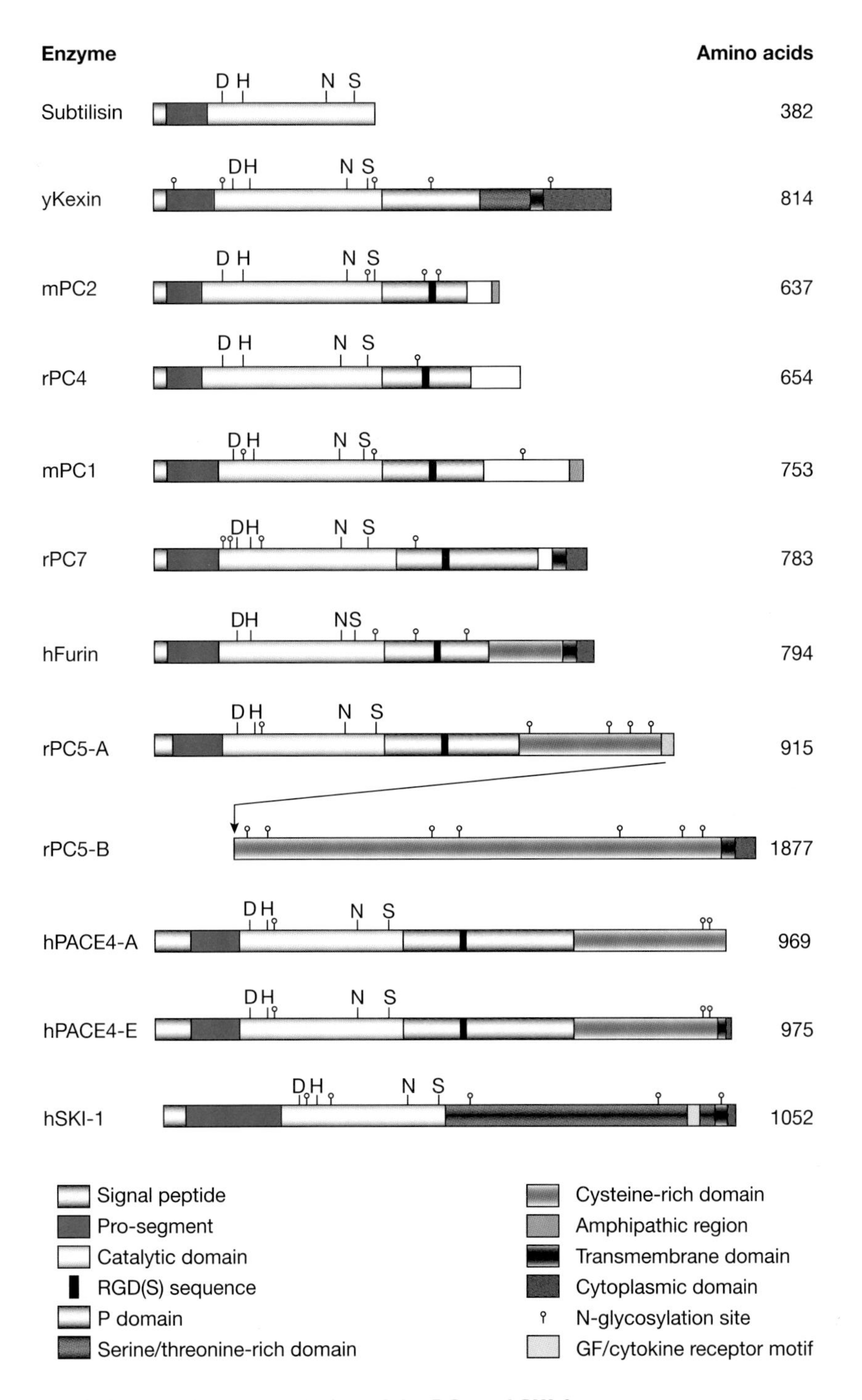


**Figure 1. Schematic representation of the PCs and SKI-1**
A comparison with bacterial subtilisin and yeast kexin is shown. The C-terminus of PACE4-E contains a hydrophobic sequence that could serve as a signal for glycosylphosphatidylinositol anchoring. y, yeast; m, mouse, r, rat; h, human; GF, growth factor. The RGD motif is present in all PCs, except for PC7, in which it is replaced by an RGS motif.

## Processing regulation within the constitutive and regulated secretory pathways

Analysis of the tissue-specific processing of substrates by PCs and SKI-1 revealed a hierarchy in the order with which these convertases cleave their cognate proproteins. Two key factors regulate this ordered process: (i) an autocatalytic zymogen activation that involves one or two cleavages resulting in the removal of the convertase pro-segment (Figure 1) [4] and (ii) the subcellular localization of each enzyme (Figure 2).

### Autocatalytic activation of the PCs

Analysis of the biosynthesis of furin, PC1, PC2, PACE4, PC5, PC7, and SKI-1 [1–5] revealed that their N-terminal pro-segment acts as an intramolecular chaperone and, with the exception of PC2 and SKI-1, as a nanomolar inhibitor of its cognate enzyme. Indeed, overexpression of the pro-segments of furin, PC7 and PC5 as independent domains confirmed their inhibitory potency and revealed that the C-terminus of the pro-segment contains the critical inhibitory

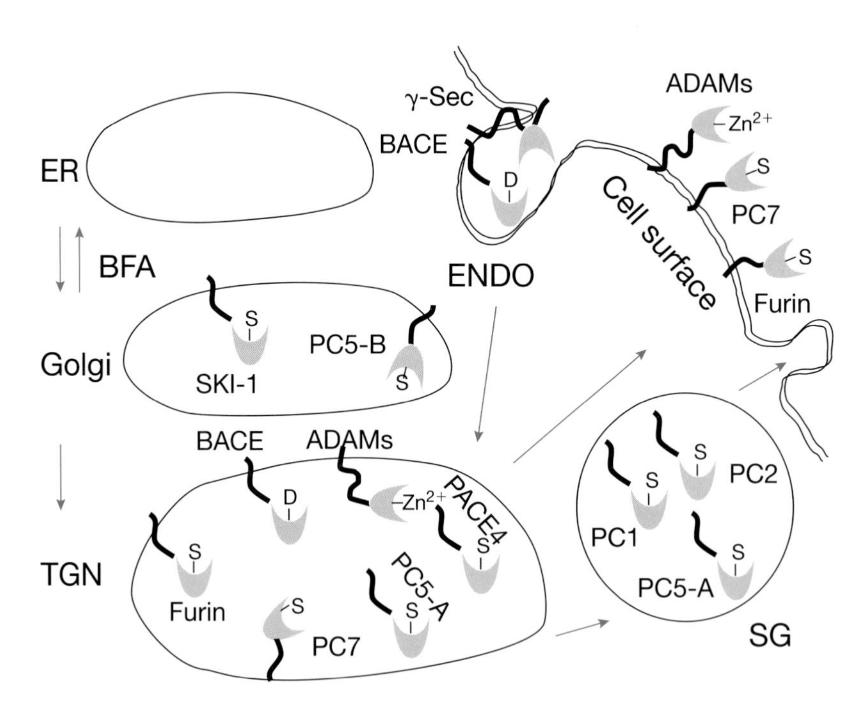


**Figure 2. Subcellular localization of PCs, SKI-1, ADAMs, BACE and the presumptive γ-secretase**

Both SKI-1 and PC5-B are localized in Golgi saccules that fuse with the ER upon treatment with brefeldin A (BFA). Furin, PACE4, PC5-A and PC7, BACE (β-site amyloid-β protein precursor-cleaving enzyme) and ADAMs (a disintegrin and metalloproteinase) are concentrated in the TGN, and the membrane-bound enzymes cycle to the cell surface and back via endosomes (ENDO; possibly containing γ-secretase). In contrast, PC1, PC2 and some PC5-A enter regulated secretory granules (SG). Enzymes that contain membrane-spanning domains are emphasized with bold dark stalks.

**Table 1. Alignment of the processing sites of SKI-1 substrates**

| Precursor protein | Cleavage site sequence | | | | | | | | | | | | | | | | |
|---|---|---|---|---|---|---|---|---|---|---|---|---|---|---|---|---|---|
| | **P8** | **P7** | **P6** | **P5** | **P4** | **P3** | **P2** | **P1** | ↓ | **P1′** | **P2′** | **P3′** | **P4′** | **P5′** | **P6′** | **P7′** | **P8′** |
| (h)proSKI-1/S1P | | | | | | | | | | | | | | | | | |
| Site B | Arg | Lys | Val | Phe | **Arg** | Ser | **Leu** | Lys | ↓ | Tyr | Ala | Glu | Ser | Asp | Pro | Thr | Val |
| Site B′ | Val | Thr | Pro | Gln | **Arg** | Lys | **Val** | Phe | ↓ | Arg | Ser | Leu | Lys | Tyr | Ala | Glu | Ser |
| Site C | Arg | His | Ser | Ser | **Arg** | Arg | **Leu** | Leu | ↓ | Arg | Ala | Ile | Pro | Arg | Gln | Val | Ala |
| (h)SREBP-2 | Ser | Gly | Ser | Gly | **Arg** | Ser | **Val** | Leu | ↓ | Ser | Phe | Glu | Ser | Gly | Ser | Gly | Gly |
| (h)SREBP-1a | His | Ser | Pro | Gly | **Arg** | Asn | **Val** | Leu | ↓ | Gly | Thr | Glu | Ser | Arg | Asp | Gly | Pro |
| (h)ATF6 | Ala | Asn | Gln | Arg | **Arg** | His | **Leu** | Leu | ↓ | Gly | Phe | Ser | Ala | Lys | Glu | Ala | Gln |
| (h)proBDNF | Lys | Ala | Gly | Ser | **Arg** | Gly | **Leu** | Thr | ↓ | Ser | Leu | Ala | Asp | Thr | Phe | Glu | His |
| (r)pro-Somato-statin (antrin) | Asp | Pro | **Arg** | Leu | **Arg** | Gln | **Phe** | Leu | ↓ | Gln | Lys | Ser | Leu | Ala | Ala | Ala | Thr |

h, human; r, rat. Bold and underlined residues have been shown to be critical recognition motifs in the SKI-1 substrates.

elements. NMR analysis of this inhibitory region in PC7 revealed that it contains an α-helix that is stabilized by salt bridges and H-bonds. Expression of these pro-segments *in trans* with enzymes in which the pro-segments have been deleted (Δpro) did not generate highly active proteins in mammalian cells. In contrast, in yeast such an *in trans* complementation of Δpro-kexin with the pro-segment led to an active protease [6]. Swapping of the pro-domains of PCs demonstrated that the pro-segment of PC1 can replace that of PACE4, and the furin pro-segment can replace that of PC1 and, to a lesser extent, PC2. The latter convertase requires a specific binding protein, 7B2 [7], for its efficient zymogen activation.

**Subcellular localization of PCs**

SKI-1 is sorted to the *cis/medial* Golgi (Figure 2), which suggests that it is poised to process its cognate precursors before any other PC (Table 1). Indeed, it is in endoplasmic reticulum (ER)/early Golgi that the processing of the four known SKI-1 substrates occurs. These include the sterol regulatory element binding proteins (SREBPs), activating transcription factor 6 (ATF6), brain-derived neurotrophic factor (BDNF) and prosomatostatin (see Chapter 12 and [2]). The next convertase in line is the PC5 isoform PC5-B, that has recently been shown to localize to an earlier, brefeldin A-collapsible, saccule of the Golgi apparatus (Figure 2). Aside from being found in the *trans* Golgi network (TGN), furin and PC7 cycle from the cell surface to the TGN, a traffic controlled by their respective cytosolic tails (reviewed in [8]). Precursors processed in the constitutive pathway include most growth factors, ER-localized transcription factors, e.g. SREBPs and ATF6, adhesion molecules, surface glycoproteins and receptors such as Notch [2]. Most polypeptide prohormone processing occurs within the regulated secretory pathway and generates bioactive hormones that are secreted from mature secretory granules (SGs) in response to specific stimuli. PC1, PC2 and PC5-A (Figure 2), which concentrate in mature SGs, are the only PCs to enter this pathway, owing to the presence of specific domains, including two α-helices in the C-terminal segment of PC1 and possibly one within the last 38 amino acids of PC5-A. The cellular PC1 and PC2 activities are tightly regulated. PC1 is inhibited not only by its pro-segment, but also by the 135 amino acids of its C-termini, which contain the above-mentioned SG-targeting signals, and the novel inhibitor proSAAS [9]. The PC1-inhibitory domain within this granin-like protein has been localized [9], peptides mimicking its effect synthesized, and the critical residues identified [10]. For PC2, the binding protein/inhibitor 7B2 is necessary for both productive folding of the enzyme, as well as to temporarily inhibit its activity via its 31-amino-acid C-terminal segment (reviewed in [7]). No specific binding protein has yet been identified for PC5-A.

**Consequences of convertase inhibition or knockout**

In order to delineate the specific function of a convertase in a given cell or tissue, one needs to have methods to inhibit its expression and/or activity in a specific fashion. α-1Antitrypsin Portland (α1-PDX) is an inhibitor of the constitutive secretory pathway convertases, including furin, PACE4, PC5 and PC7 (Figure 2). We demonstrated recently that when the colon adenocarcinoma cell line HT-29 overexpresses this serpin, the cells are more prone to apoptosis and are less likely to form tumours in nude mice [11]. Preliminary data also showed that the potential of these cells to metastasize to the liver is lower than for the parental cells [12]. Pituitary AtT20 cells that overexpress α1-PDX also show an increased apoptotic phenotype and decreased growth and foci formation on soft agar. To provide insight into the genes that favour tumour formation and metastasis, we began to compare the gene expression profile of AtT20 and HT-29 cells, in presence or absence of α1-PDX, using both microarray (for mRNA) and proteomic two-dimensional gel (for protein) technologies. Microarray analysis of mouse AtT20 cell mRNA revealed the substantial indirect up- or down-regulation of approx. 18 gene products by α1-PDX. The most prominent down-regulated mRNA was that of the growth factor-inducible DNA-binding inhibitor Id3 (19-fold downregulation), which plays a vital role in the proliferation of many cell types. This suggests that blocking the processing of certain growth factors, such as platelet-derived growth factor, or their receptors may indirectly signal to down-regulate Id3. Among the major up-regulated mRNAs were thymosin β4 (42-fold upregulation) and somatostatin receptor-1 (45-fold upregulation). Thymosin β4 binds to monomeric actin and prevents spontaneous polymerization of actin monomers into filaments. It remains to be seen whether metastasis is more or less pronounced in these cells. There is still much to be learned about the consequences of inhibiting each individual PC, in terms of both mRNA and protein expression. One of the major problems encountered with the PC inhibitors is that none of them is absolutely specific for the cognate convertase, except for proSAAS (for inhibiting PC1) [9,10] and 7B2 (for inhibiting PC2) [7].

Knockout of the *PC2*, *PC4* [1–5], *PACE4* or *PC7* (E. Robertson and D. Constam, personal communication) genes resulted in viable animals with relatively mild phenotypes. In contrast, *furin* knockout mice are embryonic lethal. We have not yet been able to obtain *PC1*$^{-/-}$ mice, even though *PC1*$^{+/-}$ mice are readily born (M. Mbikay, N.G. Seidah and M. Chretien unpublished work). Thus far, no *PC5* null mice have been produced. Recently, the *SKI-1*$^{-/-}$ genotype was found to be lethal to mouse embryos and a conditional *SKI-1* knockout in liver clearly emphasized the role of this convertase in cholesterol and lipid metabolism [13]. In conclusion, the availability of single PC-null mice and/or transgenic overexpressors (and their crosses) enriches our understanding of the biological functions of PCs and provides models of pathologies that should lead ultimately to the identification of their physiological substrates.

**Role of PCs in neurodegenerative pathologies: Alzheimer's disease**

The proteins amyloid-β protein precursor (βAPP), presenilin 1 and presenilin 2 have been implicated in the early-onset autosomal dominant Alzheimer's disease (reviewed in [14]; see also Chapter 4). Mutations in these three genes account for 50% of all inherited Alzheimer's disease cases. A fourth gene that is implicated in familial Alzheimer's disease is the apolipoprotein E gene, as patients carrying its ε4 allele show an increased density of amyloid-β protein (Aβ) deposits [14]. All mutant genes that cause Alzheimer's disease increase the concentration of $A\beta_{42}$, which appears to be especially prone to forming plaques, suggesting that this process is a fundamental alteration leading to pathogenesis. Intense efforts have been directed towards the identification of the proteases involved in the processing of APP, i.e. the α-, β- and γ-secretases. The PCs seem to play an indirect role in Alzheimer's disease through the zymogen processing of both α- and β-secretases.

**α-Secretase**

Cleavage by α-secretase within the His-His-Gln-Lys$^{668}$↓Leu-Val sequence of βAPP is the major physiological route of maturation. This cleavage precludes the formation of insoluble Aβ and is therefore non-amyloidogenic. Studies aimed at identifying α-secretase(s) candidates suggest the involvement of enzymes belonging to the ADAM (a disintegrin and metalloproteinase) and PC families, including ADAM10 (kuzbanian) and ADAM17 (tumour necrosis factor-α-converting enzyme) (see Chapter 11). ADAM proteinases are first synthesized as zymogens that require activation in the TGN by as yet unidentified PC-like enzymes. We demonstrated that inhibition of PCs by α1-PDX blocks the α-secretase-mediated cleavage of βAPP, while overexpression of PC7 enhances it [15]. Furthermore, PC7 and ADAM10, but not ADAM17, are likely to contribute to the constitutive secretion of soluble APPα by human LoVo cells [16]. It is thus possible that PC7 is involved in the processing of proADAM10 either directly or indirectly. Both enzymes cycle between the TGN and the cell surface, and are thus present in compartments thought to display α-secretase activity. Thus, α-secretase activity depends on the availability of convertase(s), e.g. PC7 and possibly furin.

**β-Secretase**

The amyloidogenic pathway that generates Aβ starts with β-secretase cleaving APP at the Glu-Val-Lys-Met$^{652}$↓Asp-Ala sequence, or of its Swedish mutant form at Glu-Val-Asn-Leu$^{652}$↓Asp-Ala. This results in the generation of a membrane-bound 99-amino-acid peptide (approx. 10 kDa), known as C99. Very recently, five different groups simultaneously reported the isolation and initial characterization of the human aspartyl proteinase β-secretase (reviewed in [17]). The major β-secretase candidate is β-site APP-cleaving enzyme (BACE), a type-I membrane-bound protein with a prodomain that is rapidly cleaved intracellularly for maximal activation and cellular sorting of the proteinase. Since the processing site of the BACE pro-segment is similar to PC-cleavage sites (Figure 3), we studied the involvement of PCs in the biosynthesis of BACE as well as the molecular properties and cellular trafficking of this enzyme [18]. Our biosynthesis and microsequencing data demonstrated that furin and PC5 are the major PCs responsible for the conversion of proBACE (starting at the Thr[22]-His-Leu-Gly-Ile-Arg sequence) into BACE within the TGN by cleavage at the Arg-Leu-Pro-Arg[45]↓Glu[46]-Thr-Asp-Glu-Glu site [18]. Characteristics that are particular to BACE include palmitoylation at the three cysteine residues within its cytosolic tail and sulphation at one or more of its carbohydrate moieties. Palmitoylation of BACE may be critical for its entry into lipid rafts wherein γ-secretase is also thought to be active

### Role of SKI-1/S1P in viral infections

Thus far, SKI-1/S1P has been implicated in the processing of SREBPs, ATF6 (see also Chapter 12) and the neuropeptide precursors of BDNF and somatostastin [2]. Thus, aside from its fundamental role in cholesterol and fatty acid homoeostasis (see Chapter 12), this enzyme may also have other unsuspected functions such as cleavage of particular viral glycoproteins. Many enveloped viruses, such as influenza viruses, paramyxoviruses, HIV and filoviruses, have fusion proteins that undergo post-translational proteolytic processing by host proteases. In most cases, cleavage is an important biological control mechanism, since it triggers fusion activity and is thus essential for

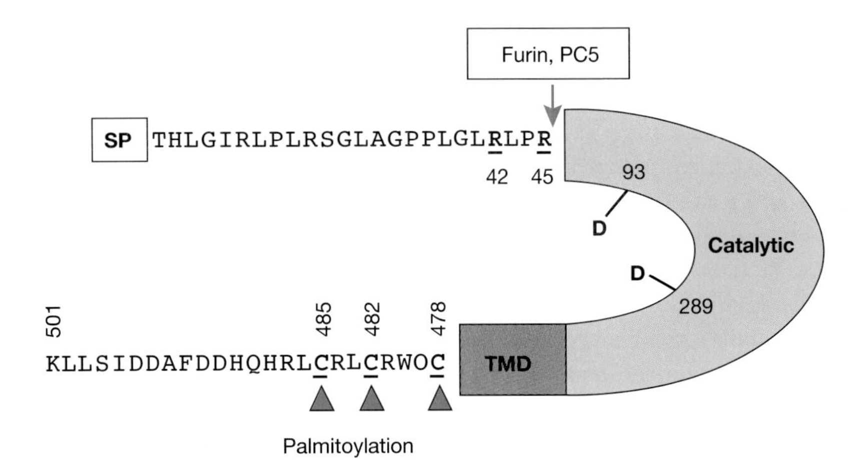


**Figure 3. Schematic representation of mouse BACE**
The signal peptide (SP), pro-segment, catalytic domain and cytosolic tail are shown. The positions of the two active-site aspartate residues (D), and the cytosolic tail palmitoylation sites (arrowheads) are emphasized. In addition, the furin/PC5-A processing sites within the pro-segment are shown. TMD, transmembrane domain.

virus entry into the host cell. Data on various viruses showed that endoproteolytic cleavage of envelope glycoproteins by one or more PCs is required for the acquisition of fusogenic potential and therefore for the infectious capacity of viral particles. As in the case of viral glycoprotein precursors, activation of some bacterial toxins also requires maturation by a convertase of the PC-family. For the HIV envelope glycoproptein, this maturation occurs within the constitutive secretory pathway where PCs such as furin, PC5-B and PC7 are localized (reviewed in [1–3]).

Lassa virus (LAV) is the causative agent of a haemorrhagic fever that is endemic in West Africa. LAV belongs to the family of Arenaviridae, which also includes lymphocytic choriomeningitis virus (LCMV), Mopeia virus and the New World arenaviruses, such as the Junin and Machupo viruses. Virions are composed of a nucleocapsid surrounded by a lipid-containing envelope and viral glycoprotein spikes. The LAV glycoprotein is synthesized as a 76-kDa precursor glycoprotein (GP-C) that is post-translationally cleaved into the N-terminal subunit GP-1 (44 kDa) and the C-terminal fragment GP-2 (36 kDa) containing the membrane anchor (Figure 4). It has been shown that LCMV and Junin virus glycoprotein cleavage occurs in the Golgi apparatus or a post-Golgi compartment. GP-1 of LCMV and of LAV interacts with a host cell-surface receptor that has been identified recently as α-dystroglycan, while the ectodomain of GP-2 contains a fusion peptide. After cleavage of the LCMV and Junin virus glycoprotein, the fusion peptide appears to be exposed by a conformational change in a pH-dependent manner to mediate fusion. Recently, the cleavage site of LAV GP-C (Arg-Arg-Leu-Leu$^{259}$↓Gly-Thr-Phe) was identified [19]. Systematic mutagenesis revealed that the recognition motif is Arg-Xaa-(Leu/Ile/Val)-Leu$^{259}$↓ and that SKI-1 is indeed the cognate convertase [20]. This led to the development of a rapid, sensitive and specific enzymic assay for SKI-1 using a quenched fluorogenic substrate that mimics the GP-C cleavage site [21] that is far superior to those reported previously [5]. The sequence conservation of cleavage sites (Figure 4) indicates that SKI-1 may cleave most Areanviridae and Bunyaviridae glycoproteins.

GP-C is the first viral glycoprotein known to be processed by SKI-1, and not by PCs. Treatment with the guanosine analogue ribavarin is the only available therapy for Lassa fever in humans, but is effective only if started very early. Specific inhibitors of SKI-1/S1P may lead to new therapeutic approaches for Lassa fever and for the devastating Arenavirus-mediated haemorraghic fever.

## NRDc: a cytosolic and cell surface dibasic convertase

The acronym NRDc reflects the preference of the enzyme for cleaving at the N-terminus of arginine residues in basic doublets. It can, however, cleave upstream of lysine residues in dibasic motifs [22]. The endopeptidase cleaves dynorphin A (at Arg↓Arg); α-neoendorphin (at Lys↓Arg) and somatostatin-28 (at ↓Arg-Lys) *in vitro* [23]. It is unlikely that these peptides constitute *bona*

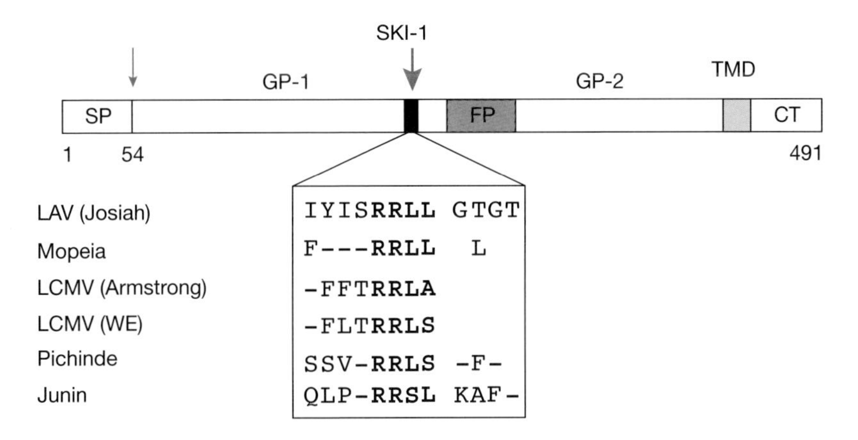


**Figure 4. Predicted processing of Arenaviridae and Bunyaviridae glycoproteins by SKI-1**

The cleavage site sequence of the envelope glycoprotein of LAV and the presumptive ones of its homologues are shown. The cleavage position is indicated by ↓. SP, signal peptide; FP, fusion peptide; TMD, transmembrane domain; CT, C-terminal. Josiah, Armstrong and WE are viral strains. Bold residues represent the recognition site.

*fide in vivo* substrates since the enzyme is present mainly in the cytosol and seems unable to transit through the secretory pathway [24]. The enzyme is also exported at the cell surface and/or in the extracellular milieu [24,25]. The cleavage specificity of NRDc suggests that it may be involved in limited intracellular or extracellular proteolysis. Thus far, none of its cytosolic and/or extracellular substrate(s) have been identified.

Two cDNAs encoding alternatively spliced forms of NRDc were isolated from cDNA libraries from rat and human testes. While the minor isoform contains an extra 68 residues close to the active site, no functional difference was observed between the two isforms [24]. Therefore NRDc will refer to the major NRDc isoform 1. The predicted rat protein (1161 amino acids; 133 kDa) shows an inverted consensus binding site for the catalytic $Zn^{2+}$, His-Xaa-Xaa-Glu-His, compared with the His-Glu-Xaa-Xaa-His sequence that is found in most metallopeptidases. This unambiguously classified the enzyme in the inverzincin/M16 family of metalloendopeptidases [26,27]. Another feature of the sequence is the presence of a 71-amino-acid acidic stretch (79% Glu+Asp), which is located approx. 30 residues upstream from the His-Xaa-Xaa-Glu-His motif (Figure 5) [23].

The closest mammalian member to NRDc is insulin-degrading enzyme (IDE; insulysin) with 36% overall identity and 47% identity in the conserved catalytic region. Several other members of the inverzincin/M16 family are involved in peptide/protein degradation as well as in peptide/protein processing, such as Axl1, which N-terminally processes the yeast cytosolic mating a-factor. Another characteristic of some members of this family, including NRDc, is their bifunctionality. For example, in addition to its protease activi-

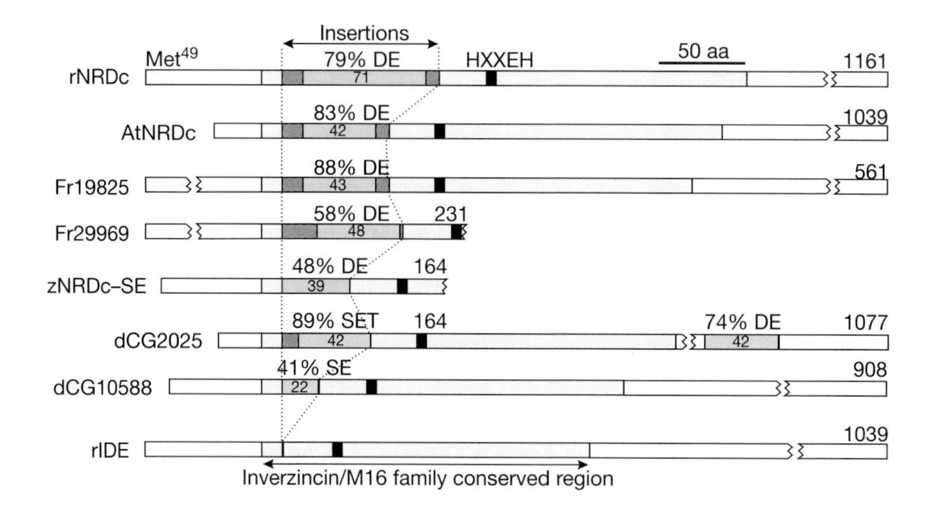


**Figure 5. Schematic alignment of NRDc-like M16 family members exhibiting one or more insertions in the catalytic region**

The conserved region of the M16 family [186 amino acids; light blue shading in the rat IDE (rIDE) sequence; lower frame] comprises the His-Xaa-Xaa-Glu-His binding motif of catalytic $Zn^{2+}$ (black). Insertion-containing members are shown in the upper frame. Length of the acidic (DE) or serine/glutamine-rich (SE) and serine/glutamine/threonine-rich (SET) stretches (dark blue shading) and the percentage content of Glu + Asp, Ser + Glu + Asp and Ser + Glu + Asp + Thr are given for the DE, SE and SET domains respectively. These stretches are either flanked or not by short sequences (grey) that are non-homologous with the M16 family consensus. Note that in addition to its SET domain, dCG2025 possesses an acidic stretch located approx. 50 amino acids upstream of its C-terminus. rNRDc, rat NRDc (NP037125); AtNRDc, *A. thaliana* putative NRDc (AAF63132); Fr19825, *F. rubripes* (JGI 19825); Fr29969, *F. rubripes* (JGI 29969); zNRDc–SE, *Danio rerio* (AW342949); dCG2025, *D. melanogaster* (AAF48105); dCG10588, *D. melanogaster* (AAF51661); rIDE (S29509).

ty, Axl1 shows non-enzymic functions in yeast budding. Alignment of M16 family members reveals an approx. 200-amino-acid conserved region comprising the His-Xaa-Xaa-Glu-His motif (Figure 5).

## NRDc acidic domain

NRDc was the first member of the family shown to have an insertion (90 amino acids) in this region, which is composed of its acidic stretch extended by 12 N- terminal and seven C-terminal residues (acidic domain; DAC). We recently identified, in data banks, six members presenting acidic or non-acidic insertions at exactly the same position (Figure 5). While the proteins from *Arabidopsis thaliana* and *Fugu rubripes* (the Japanese puffer fish) also show acidic insertions, those from zebrafish and *Drosophila melanogaster* are serine- and glutamine-rich. The latter comprise phosphorylation sites and may thus be more negatively charged than expected from the number of glutamine residues. Interestingly, in addition to its SET (serine-, glutamine- and threonine-rich) domain, *D. melanogaster* CG2025 also shows an acidic stretch at its

C-terminus. With 82–87% identity to NRDc, the *F. rubripes* and zebrafish sequences are probably NRDc orthologues. On the other hand, the *D. melanogaster* sequences are only 14% and 4% more identical to NRDc and IDE respectively, while the *A. thaliana* protein is even slightly more related to IDE (48%) than to NRDc (46%). Such domains may thus have been acquired or remodelled at different stages during evolution. What is their function(s)? In NRDc, the 90-amino-acid acidic domain is not required for catalytic activity, suggesting that it constitutes an appendix to the global enzyme structure that is not required for its folding [28].

### Cellular and subcellular distribution

NRDc transcripts are particularly abundant in the testes, skeletal muscle and heart. During mouse development, NRDc transcripts appear first in the central nervous system and cephalic and spinal ganglia (embryonic day 10.5), and later on in many other tissues [29]. In the testes, NRDc is only detected in germ cells and is associated with the axoneme of the flagellum [25]. Although it is mainly a cytosolic protein, NRDc is also secreted by germ cells and by various cell lines, and can be retained at the cell surface, though it is devoid of a transmembrane domain. This intracellular/extracellular localization of NRDc, also observed for IDE and the metalloendopeptidases 24.15 and 24.16, raises the question of the mechanism of its export, which is still not known.

### NRDc modulates the heparin-binding epidermal growth factor (EGF)-like growth factor (HB-EGF) cellular response and is potently inhibited by HB-EGF

In addition to its peptidase activity, NRDc binds with a high affinity to HB-EGF at the cell surface ($K_d = 0.5–5$ nM) and regulates its activity [30]. HB-EGF is a potent stimulator of cell proliferation and migration, and is implicated in various physiological processes such as wound healing, blastocyst implantation, smooth muscle cell hyperplasia, atherosclerosis and tumour growth. NRDc levels, modulated by overexpression and antisense Morpholino oligomers, appear to correlate with the observed enhancement of the HB-EGF-induced response [30]. Although the molecular mechanism of this potentiation remains to be elucidated, NRDc may promote the interaction of soluble HB-EGF with its receptor. In addition, its enzyme activity is not required since an active site mutant enhances HB-EGF-induced migration to the same degree as the wild-type enzyme [30]. It is noteworthy that NRDc does not bind to other growth factors of the EGF family and is thus specific for HB-EGF. While NRDc modulates the cellular response induced by HB-EGF, the latter in turn regulates NRDc enzyme activity, since it is a potent inhibitor with an $IC_{50}$ of approx. 100 nM. The NRDc–HB-EGF interaction relies on the 21-amino-acid basic heparin-binding domain of HB-EGF and the acidic domain of NRDc [28]. An NRDc mutant that lacks this domain is approx. 6-fold less sensitive to HB-EGF and when expressed at the cell surface, does not bind HB-EGF.

Finally, through its acidic domain, NRDc's enzymic activity as well as its HB-EGF binding may be regulated by extracellular $Ca^{2+}$.

## Conclusions

Within the last decade the longstanding goal of identifying the convertases involved in the limited proteolysis of secretory proteins has been largely achieved. The identification of the seven kexin-like PCs and the pyrolysin-like SKI-1/S1P has paved the way to major advances in our understanding of the proteolytic machinery that regulates multiple homoeostatic and pathological events. While these secretory processes are now better understood, the study of similar cleavages occurring within the cytosol and/or at the cell surface is in an explosive phase. NRDc cleavage specificity and the presence of the enzyme in both cytosol and cell surface suggest that it may be an equivalent to the PCs in these compartments.

### Summary


- *Eight mammalian PCs related to the bacterial serine protease subtilisin are known. PC1, PC2, furin, PACE4, PC4, PC5 and PC7 process precursors at the C-terminus of single or paired basic residues. In contrast, SKI-1/S1P cleaves proproteins at hydrophobic residues.*
- *PC1 and PC2 are responsible for the processing of most endocrine and neuropeptide precursors, including a large variety of proteins sorted to secretory granules. The other PCs process proteins that traffic though the constitutive secretory pathway. SKI-1/S1P cleaves precursors within the cis/medial Golgi, including growth factors and membrane-bound transcription factors.*
- *Over-expression of PCs leads to tumour formation and metastasis, as well as increased angiogenesis. As a corollary, inhibition or genetic silencing of PCs or SKI-1 leads to decreased tumour formation and metastasis, and lowers the infectivity of a number of pathogenic viruses including HIV and Lassa virus.*
- *The PCs are responsible for the processing and activation of a large number of proproteins, including the secretase enzymes that are implicated in neurodegenerative pathologies such as Alzheimer's disease.*
- *In the cytosol, nardilysin or NRDc represents a metallopeptidase that cleaves polypeptides at the N-terminus of paired basic, arginine and lysine, residues. It is also secreted in the extracellular milieu and therein it binds and is inhibited by the oncogenic polypeptide HB-EGF.*


We thank Brigitte Mary for secretarial help. This work was supported by a Canadian Institutes of Health Research group grant GPC 11474.

## References


1. Seidah, N.G. & Chretien, M. (1997) Eukaryotic protein processing: endoproteolysis of precursor proteins. *Curr. Opin. Biotechnol.* **8**, 602–607
2. Seidah, N.G. & Chretien, M. (1999) Proprotein and prohormone convertases: a family of subtilases generating diverse bioactive polypeptides. *Brain Res.* **848**, 45–62
3. Seidah, N.G. (2001) Cellular limited proteolysis of precursor proteins and peptides, in *The Enzymes: Volume XXII Co- and Posttranslational Proteolysis of Proteins* (Dalbey R.E. & Sigman D.S., eds), pp. 237–258, Academic Press, New York
4. Steiner, D.F. (2001) The prohormone convertases and precursor processing in protein biosynthesis, in *The Enzymes: Volume XXII Co- and Posttranslational Proteolysis of Proteins* (Dalbey R.E. & Sigman D.S., eds), pp. 164–197, Academic Press, New York
5. Elagoz, A., Benjannet, S., Mammarbassi, A., Wickham, L. & Seidah N.G. (2002) Biosynthesis and cellular trafficking of the convertase SKI-1/S1P: Ecotodomain shedding requires SKI-1 activity. *J. Biol. Chem.* **277**, 11265–11275
6. Lesage, G., Prat, A., Lacombe, J., Thomas, D.Y., Seidah, N.G. & Boileau, G. (2000) The Kex2p proregion is essential for the biosynthesis of an active enzyme and requires a C-terminal basic residue for its function. *Mol. Biol. Cell* **11**, 1947–1957
7. Mbikay, M., Seidah, N.G. & Chretien, M. (2001) Neuroendocrine secretory protein 7B2: structure, expression and functions. *Biochem. J.* **357**, 329–342
8. Molloy, S.S., Anderson, E.D., Jean, F. & Thomas, G. (1999) Bi-cycling the furin pathway: from TGN localization to pathogen activation and embryogenesis. *Trends Cell Biol.* **9**, 28–35
9. Fricker, L.D., McKinzie, A.A., Sun, J., Curran, E., Qian, Y., Yan, L., Patterson, S.D., Courchesne, P.L., Richards, B., Levin, N. et al. (2000) Identification and characterization of proSAAS, a granin-like neuroendocrine peptide precursor that inhibits prohormone processing. *J. Neurosci.* **20**, 639–648
10. Basak, A., Koch, P., Dupelle, M., Fricker, L.D., Devi, L.A., Chretien, M. & Seidah, N.G. (2001) Inhibitory specificity and potency of proSAAS-derived peptides toward proprotein convertase 1. *J. Biol. Chem.* **276**, 32720–32728
11. Khatib, A.-M., Siegfried, G., Prat, A., Luis, J., Chrétien, M., Metrakos P. & Seidah, N.G. (2001) Inhibition of proprotein convertases is associated with loss of growth and tumorigenicity of HT-29 human colon carcinoma cells: importance of IGF-1R processing in IGF1-mediated functions. *J. Biol. Chem.* **276**, 30686–30693
12. Khatib, A.-M., Siegfried, G., Chretien, M., Metrakos, P. & Seidah, N.G. (2002) Proprotein convertases in tumor progression and malignancy: novel targets in cancer therapy. *Am. J. Pathol.* **160**, 1921–1935
13. Yang, J., Goldstein, J.L. Hammer, R.E. Moon, Y.A. Brown, M.S. & Horton, J.D. (2001) Decreased lipid synthesis in livers of mice with disrupted site-1 protease gene. *Proc. Natl. Acad. Sci. U.S.A.* **98**, 13607–13612
14. St George-Hyslop, P.H. (2000) Molecular genetics of Alzheimer disease [Review]. *Semin. Neurol.* **19**, 371–383
15. Lopez-Perez, E., Seidah, N.G. & Checler, F. (1999) Proprotein convertase activity contributes to the processing of the Alzheimer's beta-amyloid precursor protein in human cells: evidence for a role of the prohormone convertase PC7 in the constitutive alpha-secretase pathway. *J. Neurochem.* **73**, 2056–2062
16. Lopez-Perez, E., Zhang, Y., Frank, S., Creemers, J., Seidah, N.G. & Checler, F. (2001) Constitutive α-secretase cleavage of the β-amyloid precursor protein in the furin-deficient lovo cell line: Involvement of the prohormone convertase 7 (PC7) and the disintegrin metalloprotease ADAM10. *J. Neurochem.* **76**, 1532–1539
17. Vassar, R. (2001) The beta-secretase, BACE: a prime drug target for Alzheimer's disease. *J. Mol Neurosci* **17**, 157–170

18. Benjannet, S., Elagoz, A., Wickham, L., Mammarbassi, A., Munzer, J.S., Basak, A., Lazure, C., Cromlish, J., Sosodia, S., Checler, F. et al. (2001) Post-translational processing of β-secretase (BACE) and its ectodomain shedding: the pro- and transmembrane/cytosolic domains affect its activity and amyloid Aβ production. *J. Biol. Chem.* **276**, 10879–10887
19. Lenz, O., ter Meulen, J., Feldmann, H., Klenk, H.D. and Garten, W. (2000) Identification of a novel consensus sequence at the cleavage site of the Lassa virus glycoprotein. *J. Virol.* **74**, 11418–11421
20. Lenz, O., ter Meulen, J., Klenk, H.D., Seidah, N.G. & Garten, W. (2001) Cleavage of GP-C of Lassa virus by subtilase SKI-1/S1P, a novel type of viral glycoprotein processing. *Proc. Natl. Acad. Sci. U.S.A.* **98**, 12701–12705
21. Basak, A., Chretien, M. & Seidah N.G. (2002) A rapid fluorometric assay for the proteloytic activity of SKI-1/S1P based on the surface glycoprotein of the hemorrhagic fever Lassa virus. *FEBS Lett.* **514**, 333–339
22. Chow, K.M., Csuhai, E., Juliano, M.A., St Pyrek, J., Juliano, L. & Hersh, L.B. (2000) Studies on the subsite specificity of rat nardilysin (N-arginine dibasic convertase). *J. Biol. Chem.* **275**, 19545–19551
23. Cohen, P., Pierotti, A.R., Chesneau, V., Foulon, T. & Prat, A. (1995) N-arginine dibasic convertase. *Methods Enzymol.* **248**, 703–716
24. Hospital, V., Chesneau, V., Balogh, A., Joulie, C., Seidah, N.G., Cohen, P. & Prat, A. (2000) N-arginine dibasic convertase (nardilysin) isoforms are soluble dibasic-specific metalloendopeptidases that localize in the cytoplasm and at the cell surface. *Biochem. J.* **349**, 587–597
25. Chesneau, V., Prat, A., Segretain, D., Hospital, V., Dupaix, A., Foulon, T., Jegou, B. & Cohen, P. (1996) NRD convertase: a putative processing endopeptidase associated with the axoneme and the manchette in late spermatids. *J. Cell Sci.* **109**, 2737–2745
26. Hooper, N.M. (1994) Families of zinc metalloproteases. *FEBS Lett.* **354**, 1–6
27. Barrett, A.J. (1998) Introduction: clan ME containing pitrilysin and its relatives. In *Handbook of Proteolytic Enzymes* (Barrett, A.J., Rawlings, N.D. & Woessner, J.F., eds), pp. 1360–1379, Academic Press, London
28. Hospital, W., Nisahi, E., Klagsbrun, M., Cohen, P., Seidah, N.G. & Prat, A. (2002) The metalloendopeptidase nardilysin (NRDc) is potently inhibited by heparin-binding EGF-like growth factor (HB-EGF). *Biochem. J.*, doi: 10.1042/BJ20020822
29. Fumagalli, P., Accarino, M., Egeo, A., Scartezzini, P., Rappazzo, G., Pizzuti, A., Avvantaggiato, V., Simeone, A., Arrigo, G., Zuffardi, O. et al. (1998) Human NRD convertase: a highly conserved metalloendopeptidase expressed at specific sites during development and in adult tissues. *Genomics* **47**, 238–245
30. Nishi, E., Prat, A., Hospital, V., Elenius, K. & Klagsbrun, M. (2001) N-arginine dibasic convertase is a specific receptor for heparin-binding EGF-like growth factor that mediates cell migration. *EMBO J.* **20**, 3342–3350

# 8

# Proteases in blood clotting


## Peter N. Walsh[1] and Syed S. Ahmad

*Sol Sherry Thrombosis Research Center, Temple University School of Medicine, 3400 North Broad Street, Philadelphia, PA 19140, U.S.A., Department of Medicine, Temple University School of Medicine, 3400 North Broad Street, Philadelphia, PA 19140, U.S.A., Department of Medicine, Temple University School of Medicine, 3400 North Broad Street, Philadelphia, PA 19140, U.S.A., and Department of Biochemistry, Temple University School of Medicine, 3400 North Broad Street, Philadelphia, PA 19140, U.S.A.*



## Abstract

The serine proteases, cofactors and cell-receptor molecules that comprise the haemostatic mechanism are highly conserved modular proteins that have evolved to participate in biochemical reactions in blood coagulation, anticoagulation and fibrinolysis. Blood coagulation is initiated by exposure of tissue factor, which forms a complex with factor VIIa and factor X, which results in the generation of small quantities of thrombin and is rapidly shutdown by the tissue factor pathway inhibitor. The generation of these small quantities of thrombin then activates factor XI, resulting in a sequence of events that lead to the activation of factor IX, factor X and prothrombin. Sufficient thrombin is generated to effect normal haemostasis by converting fibrinogen into fibrin. The anticoagulant pathways that regulate blood coagulation include the protein C anticoagulant mechanism, the serine protease inhibitors in plasma, and the Kunitz-like inhibitors, tissue factor pathway inhibitor and protease nexin 2. Finally, the fibrinolytic mechanism



[1]*To whom correspondence should be addressed at the Sol Sherry Thrombosis Research Center, Temple University School of Medicine, 3400 North Broad Street, Philadelphia, PA 19140, U.S.A. (e-mail: pnw@astro.temple.edu).*

that comprises the activation of plasminogen into plasmin prevents excessive fibrin accumulation by promoting local dissolution of thrombi and promoting wound healing by reestablishment of blood flow.

## Introduction

Virtually all complex multicellular organisms with highly developed cardiovascular systems have evolved tightly regulated mechanisms of haemostasis, which is defined as the maintenance of the fluidity of circulating blood while at the same time protecting the organism from life-threatening bleeding at sites of vascular injury, particularly in the high-pressure, high-flow vascular systems that are characteristic of mammals. This cell-mediated system of proteolysis of plasma coagulation proteins is essential to the survival of the organism as evidenced by the severe bleeding complications [1,2] that can arise in patients with haemophilia (e.g. resulting from a deficiency in clotting factors VIII or IX). However, pathological consequences can also arise from a deficiency in or abnormal regulation of anticoagulant proteases and their cofactors (e.g. deficiency in protein C, protein S, antithrombin III or factor V-Leiden), resulting in arterial or venous thrombosis [3,4].

The haemostatic mechanism has evolved as a complex system in which serine proteases with highly developed specificity assemble on cell membrane receptors in complex with cofactor molecules. They then propagate a highly regulated series of enzymic reactions that lead to the formation of a haemostatic thrombus that arrests bleeding. Haemostasis comprises a closely interwoven network of reactions in response to vascular injury. It can, perhaps artificially, be divided into the processes of platelet plug formation, blood coagulation, anticoagulation and fibrinolysis. Primary haemostasis or platelet plug formation comprises a series of events in which platelets initially adhere to subendothelial components that are exposed upon vascular injury; they then aggregate to form a haemostatic thrombus. Fibrinolysis comprises the system of proteases and activators that dissolve fibrin clots that are formed when thrombin is generated as a consequence of the blood coagulation mechanisms (Figure 1). The processes of blood coagulation, anticoagulation and fibrinolysis are all regulated by naturally occurring inhibitory molecules including serine protease inhibitors (or SERPINs) and Kunitz-type protease inhibitors (or kunins). Accumulating evidence supports the hypothesis [5] that blood coagulation is a cell surface receptor-mediated process involving highly specialized anucleate cell fragments (i.e. platelets) whose sole function is surveillance of the vascular system for breaches of its integrity. At sites of vascular injury, platelets adhere to exposed subendothelial components, become activated, aggregate to form haemostatic thrombi and expose receptors that promote the assembly of blood coagulation complexes, which leads to a massive generation of fibrin and arrest of haemorrhage [6,7], as shown schematically in Figure 1. The purpose of this article is to review the recent progress and future perspectives on the structure, function and phys-

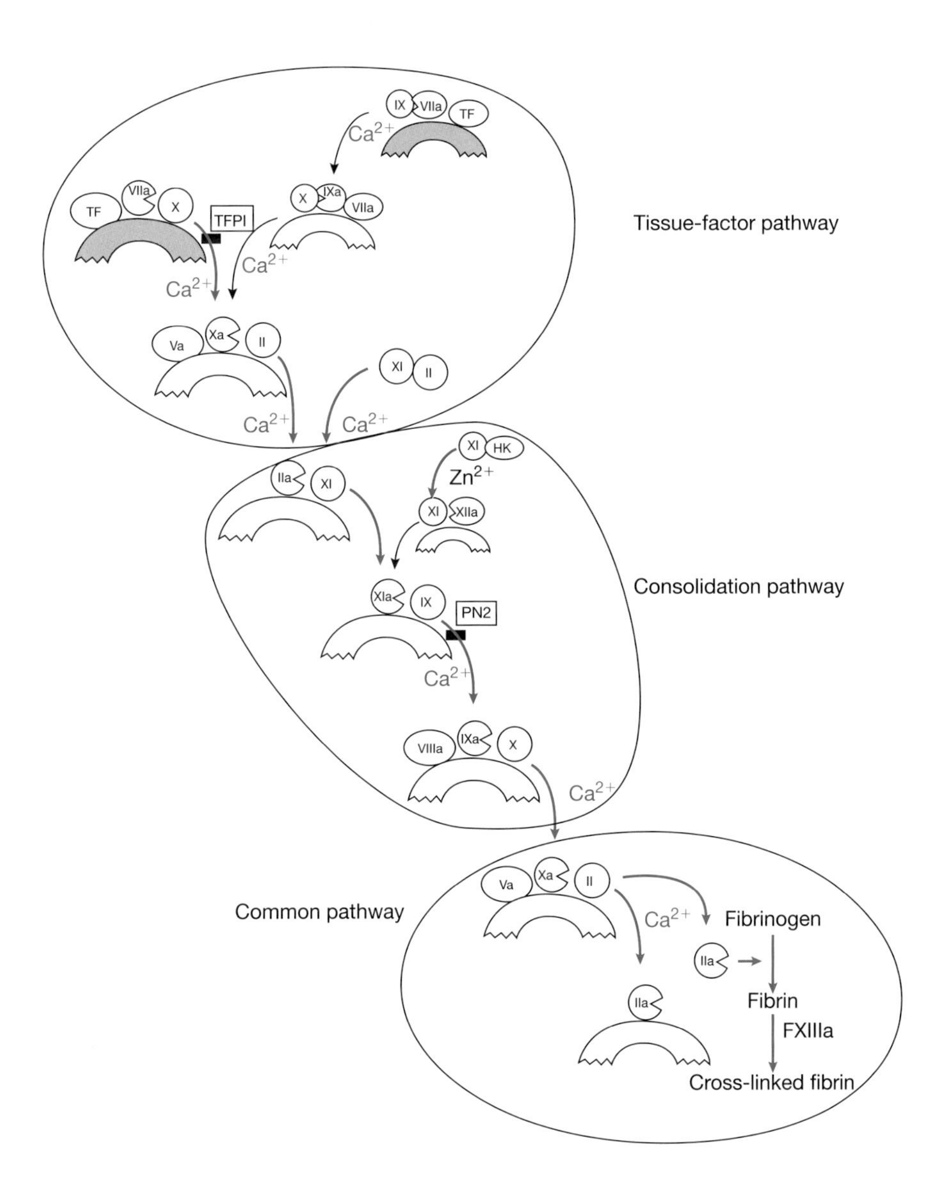


**Figure 1. The postulated sequence of blood coagulation reactions**
The tissue factor–factor VIIa complex is assembled on tissue-factor-bearing cells. The remainder of the complexes are assembled on the activated platelet membrane. The Roman numerals represent coagulation proteins in the zymogen or cofactor form, with the 'a' representing the active enzyme or cofactor. Circles represent zymogens; circles with sectors removed represent active enzymes; ellipses represent cofactors; and rectangles represent Kunitz-type inhibitors. The arrows represent conversions from zymogens to active enzymes. Fibrin is covalently crosslinked by the transglutaminase, factor XIIIa. The black arrows indicate reactions that are less physiologically relevant or secondary pathways (e.g. the secondary role of the contact proteins in and the contribution of tissue factor–factor VIIa to factor IX activation). II, prothrombin; IIa, thrombin; TF, tissue factor. See text for detailed discussion. Adapted from Walsh, P.N. (2001) Role of platelets and factor XI in the initiation of blood coagulation by thrombin. *Thromb. Haemostasis* **86**, 75–82, with permission from Schattauer GmbH.

iological roles of procoagulant, anticoagulant and fibrinolytic proteases, their regulation and their roles in the haemostatic mechanism.

## Pathways and proteases in haemostasis

Numerous serine proteases, cofactor molecules and cell surface receptors have been identified whose biological functions are essential for the initiation and propagation of procoagulant, anticoagulant and fibrinolytic responses to vascular injury. These are essential both for maintaining the fluidity of circulating blood and for effecting normal haemostasis. The biochemical properties and concentrations in plasma of these proteins are summarized in

**Table 1. Properties of proteases and cofactors in blood coagulation and fibrinolysis**

Pan modular refers to proteins containing a domain homologous with the Apple domains of factor XI and prekallikrein [17]. Gla-containing refers to the vitamin K-dependent family of proteins that contain the post-translationally generated amino acid, $\gamma$-carboxyglutamic acid. ND, not determined.

| Protease | Molecular mass (kDa) | Plasma concentration | | Plasma half-life (h) |
|---|---|---|---|---|
| | | mg/l | nM | |
| Pan modular | | | | |
| PK | 86 | 50 | 400 | 90 |
| Factor XI | 160 | 5 | 30 | 80 |
| Gla-containing | | | | |
| Factor VII | 50 | 0.5 | 10 | 5 |
| Factor IX | 75 | 5 | 90 | 24 |
| Factor X | 78 | 8 | 170 | 40 |
| Factor II | 72 | 100 | 1400 | 72 |
| Protein C | 62 | 4 | 64 | 10 |
| Protein S | 69 | 25 | 300 | ND |
| Protein Z | 62 | 2.9 | 36 | 60 |
| Other proteases | | | | |
| Factor XII | 80 | 30 | 400 | 52 |
| Plasminogen | 90 | 120 | 2100 | 52 |
| Urokinase-type plasminogen activator | 52 | 0.005 | 0.04 | 0.17 |
| t-PA | 68 | 0.005 | 0.05 | 0.08 |
| Cofactors | | | | |
| Factor V | 330 | 10 | 30 | 36 |
| Factor VIII | 330 | 0.1 | 0.3 | 12 |
| HK | 110 | 70 | 600 | 120 |

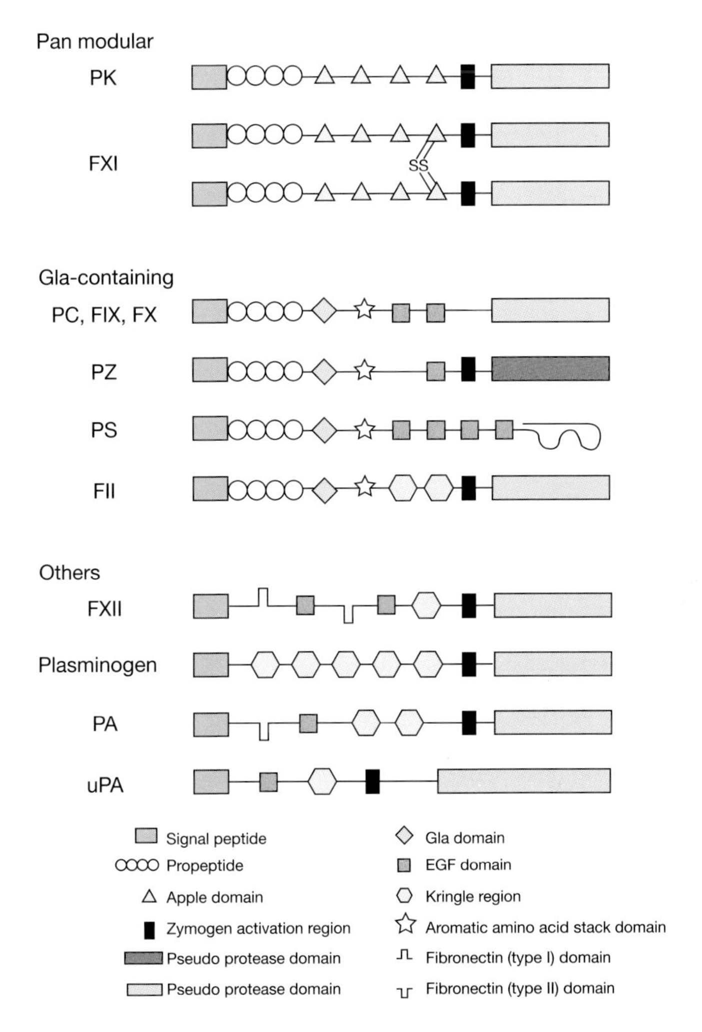


**Figure 2. Schematic representation of the nascent forms of proteases involved in blood coagulation and fibrinolysis**
Proteins are grouped as pan modular [PK and factor XI (FXI)], Gla-containing [factor IX (FIX), factor X (FX), protein C (PC), protein Z (PZ), protein S (PS) and prothrombin (FII)] and as others [factor XII (FXII), plasminogen, t-PA and urokinase-type plasminogen activator (uPA)]. Pan modular refers to proteins containing domains homologous with the Apple domains of factor XI and prekallikrein [14]. Gla-containing refers to the vitamin-K-dependent family of proteins that contain the post-translationally generated amino acid γ-carboxyglutamic acid. EGF, epidermal growth factor.

Table 1 and the domain structures are presented in schematic form in Figures 2 and 3. The serine proteases involved in coagulation, anticoagulation and fibrinolysis have all evolved from a common trypsin-like ancestral protease [8,9], and are characterized by the presence of highly conserved catalytic residues ($Ser^{195}$, $His^{57}$ and $Asp^{102}$; chymotrypsin numbering system) that are referred to as the catalytic triad, as well as modular protein domains that are highly

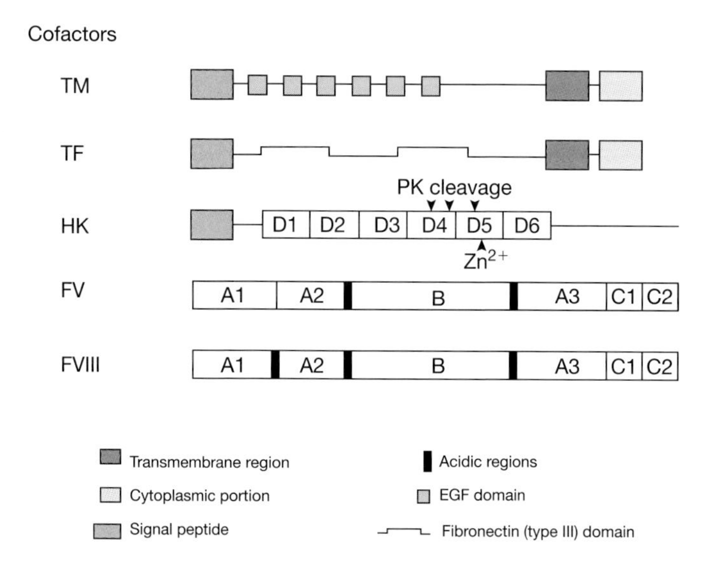


**Figure 3. Schematic representation of cofactors of blood coagulation**
TM, thrombomodulin; TF, tissue factor; FV, factor V; FVIII, factor VIII; EGF, epidermal growth factor. The various domains are identified in the key.

conserved three-dimensional structures serving similar functions in distinct, evolutionarily related protein families (Figure 2). These include the γ-carboxyglutamic acid domains that contain vitamin-K-dependent, post-translationally modified glutamic acid residues that bind calcium, thereby facilitating interactions of these proteins with biological membranes. Other protein modules (e.g. epidermal growth factor domains, Apple domains and Kringle domains) have highly conserved tertiary structures, but have evolved to promote specific protein–protein interactions that involve various proteases (Figure 2) and cofactor molecules (Figure 3). The cofactor and cell receptor molecules that participate in the assembly of enzymic complexes and enhance their catalytic efficiency are also members of evolutionarily related families that are assembled from protein modules [8] as shown in Figure 3.

Theories concerning the sequence of biochemical events comprising the haemostatic response have evolved over time as knowledge concerning the identification and functional properties of the various proteases, cofactors, cell receptor molecules and regulatory proteins has increased steadily. The first modern comprehensive theories of blood coagulation [10,11] emphasized the biological amplification that is inherent in the sequential activation of coagulation proteins, each of which converts its substrate into an active serine protease that then serves as the enzyme in the next reaction in the sequence; this results ultimately in the local explosive generation of fibrin from fibrinogen at the site of vascular injury. These and subsequent theories emphasized the so-called extrinsic and intrinsic pathways of blood coagulation as separate, distinct, and alternative mechanisms. Recent revisions to these concepts emphasized not only an integrated network of enzymic reactions [12,13], but also the fact that virtually all the enzymic reactions of blood coagulation, anticoagulation and fibrinolysis occur on cell surface membrane receptors that co-localize enzymes, cofactors and substrates in a kinetically favourable complex, the assembly of which requires initial cellular activation [5,7,14]. The following discussion considers these reaction pathways in the order in which they are thought to occur on the biological membrane regarded as the physiological locus of haemostatic reactions (i.e. the activated platelet), as opposed to the vascular endothelium, which not only contains the blood, but also maintains it in a fluid state by exerting its anticoagulant properties. These concepts are presented in schematic form in Figure 1 as the postulated sequence on cell membranes of enzymic reactions that constitute blood coagulation.

## Initiation of blood coagulation

The early theories of blood coagulation regarded the intrinsic and extrinsic pathways of factor X activation as comprising alternative mechanisms for thrombin generation and the conversion of fibrinogen into fibrin [10,11]. The intrinsic pathway was regarded as being triggered by the reciprocal proteolytic activation of the so-called contact factors [15], factor XII, high molecular weight kininogen (HK) and prekallikrein (PK) on negatively charged surfaces, which leads to the conversion of inactive factor XI into active factor XIa and the subsequent sequential activation of factor IX, factor X and prothrombin. Alternatively, exposure of tissue factor at sites of vascular injury leads to the formation of a complex with factor VIIa [16], which activates factor X, which then converts prothrombin into active thrombin in the presence of factor V and a suitable surface with exposed negatively charged aminophospholipids, in particular, platelets. These concepts have been re-examined in the light of evidence that deficiencies of the so-called contact proteins, factor XII, HK, and PK, are virtually never associated with haemostatic deficiency or for that matter any abnormal phenotype [15]. Therefore, the physiological role of the contact phase in the initiation of blood coagulation has been challenged, and this has resulted in a search for alternative mechanisms to explain the initiation of blood coagulation. In contrast, incontrovertible evidence of the importance of the intrinsic pathway for maintenance of normal haemostasis is the severe bleeding phenotype [2] observed in patients with haemophilia A (factor VIII deficiency) and haemophilia B (factor IX deficiency) as well as the milder but still clinically significant bleeding manifestations of haemophilia C (factor XI deficiency). The facts that human factor VII deficiency is associated with a rather mild bleeding tendency and that murine knockouts of the genes for factor VII and for tissue factor are associated with severe bleeding complications has focused attention on the tissue factor/factor VII pathway as the most physiologically relevant initiating mechanism in blood coagulation [16].

### Tissue factor (initiation) pathway

Tissue factor is an integral membrane protein that is normally present on the surface of certain cell types that are mainly localized outside the vasculature, whereas the soluble proteins that comprise the blood coagulation cascade are found in the blood. Vascular injury results in the exposure of tissue factor, which then binds factor VII and factor VIIa, the enzymically active form that arises after cleavage of a single peptide bond [16]. Recent evidence supports the view that circulating blood cells and endothelial cells, which do not normally express tissue factor constitutively, can acquire tissue factor activity within their membranes after interaction with subendothelial tissues and circulating microparticles that do express tissue factor [17]. These observations have given rise to the novel hypothesis that blood cells, and platelets in particular, can acquire tissue factor antigen and activity and can thereby promote the activation of factor X via the tissue factor pathway. Factor VIIa has extremely weak serine protease activity until it forms a complex with tissue factor. This complex can then activate both factor IX and factor X, with factor X being the preferred substrate for tissue factor–factor VIIa under most *in vitro* and probably *in vivo* conditions [16]. The initial generation of factor Xa molecules by the factor VIIa–tissue factor complex appears to be extremely short-lived because of the presence in plasma of a natural inhibitor, tissue factor pathway inhibitor (TFPI), a tightly binding, reversible, kunin that contains three Kunitz domains, two of which have binding sites for factor Xa, tissue factor and factor VIIa. Thus, as soon as the first molecules of factor Xa are formed, the tissue factor pathway of initiation of blood coagulation is rapidly and effectively shut down by TFPI. The factor Xa that is initially formed can also activate prothrombin in the presence of activated platelets and factor Va, but the thrombin formed by this pathway is insufficient to effect normal haemostasis.

### Consolidation pathway

Coagulation factor XI is a unique homodimeric precursor of the active serine protease, factor XIa, which is apparently the first coagulation protein within the consolidation pathway that is required for normal haemostasis, as evidenced by the bleeding complications that arise in patients with factor XI deficiency [14]. Factor XI can be activated proteolytically by three distinct serine proteases [12,13,18] (thrombin, factor XIIa, and factor XIa), all of which cleave factor XI at the $Arg^{369}$–$Ile^{370}$ peptide bond to activate the catalytic domain, which then recognizes coagulation factor IX as its normal macromolecular substrate. A negatively charged surface is required for optimal rates of factor XI activation by thrombin, factor XIIa or factor XIa. The physiologically relevant cellular membrane that binds to factor XI and co-localizes it with its cognate enzymes is the surface membrane of activated platelets, where the preferred enzyme for factor XI activation appears to be thrombin [19]. Extremely low concentrations (10–100 pM) of thrombin can activate factor XI. Although factor XI binds to thrombin-activated platelets in the presence of HK and zinc ions, where it can be activated by factor XIIa, the absence of haemostatic defects in patients with deficiencies in factor XII and HK cast some doubt upon the physiological relevance of factor XIIa-catalysed activation of factor XI, as emphasized in Figure 1. Alternatively, prothrombin can substitute for HK and calcium ions can substitute for zinc ions in the binding of factor XI to thrombin-activated platelets, and thrombin can substitute for factor XIIa to promote rapid and complete factor XI activation at low concentrations of thrombin [14,18,19]. This suggests that the physiologically relevant pathway for factor XI activation is the generation of thrombin by the tissue factor pathway, as discussed above and as shown schematically in Figure 1.

The factor XIa generated from surface-bound factor XI by thrombin or by factor XIIa can then bind with high affinity and specificity to platelet membrane receptors that co-localize it with its normal macromolecular substrate, factor IX, which also binds to specific high-affinity sites on activated platelets and leads to the rapid and efficient generation of factor IXa [14]. Alternatively, factor XIa activity can be regulated by a variety of inhibitors, including SERPINs that are present in plasma [e.g. antithrombin III, α-1-protease inhibitor, C1 inhibitor, α-2-antiplasmin, plasminogen activator inhibitor 1 (PAI-1) and protein C inhibitor] or, alternatively, by the kunin, protease nexin 2 (PN-2), a truncated form of the transmembrane Alzheimer's disease amyloid β-protein precursor, which contains a Kunitz-type serine protease inhibitor domain [14] (see Chapter 4). PN-2 is found in very low concentrations in plasma, but is secreted from platelet α-granules so that the physiological concentration of PN-2 may be brought to 3–5 nM, i.e. well above the $K_i$ for inhibition of factor XIa (300–500 pM). Since platelet-bound factor XIa appears to be protected from inactivation by PN-2, the secretion of PN-2 by platelets at sites of vascular injury may regulate solution phase factor XIa while protecting the platelet-bound protease from inactivation, thereby localizing subsequent events in coagulation to the haemostatic thrombus [20].

The next event in the sequence of enzymic reactions that comprises the blood coagulation cascade is the assembly of the factor-X-activating complex on the surface of activated platelets [6,7]. The assembly of this enzymic complex serves as a paradigm for a number of other protease–cofactor–substrate complexes that are assembled on cell membranes within the coagulation, anticoagulation and fibrinolytic pathways. The factor-X-activating complex consists of a vitamin-K-dependent serine protease (factor IXa), a cofactor molecule (factor VIIIa), and a vitamin-K-dependent substrate molecule (factor X), each of which binds to a specific, high-affinity, platelet membrane receptor. The functional consequence of this complex assembly is an enormous ($>2\times10^8$-fold) increase in catalytic efficiency ($k_{cat}/K_m$). Rate enhancements of factor X activation are directly correlated with receptor occupancy by all three components of the factor-X-activating complex, including that of a high-capacity, low-affinity shared factor X/prothrombin binding site [6,7]. Simultaneous binding and kinetic studies demonstrate co-ordinated interactions between the enzyme, the cofactor, and the substrate in the formation of the factor-X-activating complex.

### Common coagulation pathway

After the generation of factor Xa, the activated platelet surface promotes the assembly of the prothrombin-activating complex, which consists of the enzyme (factor Xa), the cofactor (factor Va), and the substrate (prothrombin). This is associated with a 200000-fold increase in the rate of prothrombin activation compared with the reaction in solution in the absence of the platelet surface and factor Va [7]. In contrast with the initial burst in the generation of factor Xa and thrombin via the tissue factor pathway, which is rapidly shut down by the action of TFPI, the second burst via the consolidation pathway is now sufficiently large to cleave the major plasma substrate for thrombin, i.e. fibrinogen. The cleavage of fibrinogen by thrombin generates a fibrin monomer which is then cross-linked by the action of factor XIIIa to form the insoluble and well-organized fibrin meshwork that forms the matrix of a blood clot [21,22], as shown in Figures 1 and 4.

## Regulatory and anticoagulant pathways in blood coagulation

Evidence continues to accumulate to support the elegance and complexity of design of the haemostatic mechanism, which has the dual function of maintaining the blood in a fluid state while simultaneously being capable of responding instantaneously to breaches in the vascular integrity that require controlled clot formation without propagating disseminated intravascular coagulation. This tight control is mediated not only by signal transduction mechanisms that lead to cell activation and exposure of receptors for coagulation proteins, but also by the complex systems of regulation that constitute the anticoagulant and inhibitor mechanisms of haemostasis.

### Protein C anticoagulant mechanism

An important natural anticoagulant mechanism [3] that limits the normal haemostatic and procoagulant response to vascular injury involves the association of the serine protease thrombin in complex with the endothelial cell surface receptor, thrombomodulin, with a substrate, protein C, which is bound to the endothelial cell protein C receptor. As shown in Figure 4, this complex assembly converts thrombin from a procoagulant molecule into an anticoagulant molecule that generates activated protein C. Activated protein C then binds to cell surfaces and platelet membranes in the presence of another cofactor molecule, protein S, with the formation of another cell surface–protease–cofactor complex whose major function is to cleave and inactivate factor Va (and also factor VIIIa) to their inactive forms, factor $V_i$ (and factor $VIII_i$). The consequence of proteolysis of factor Va (and factor VIIIa) to their inactive forms is a rapid and effective curtailment of further thrombin

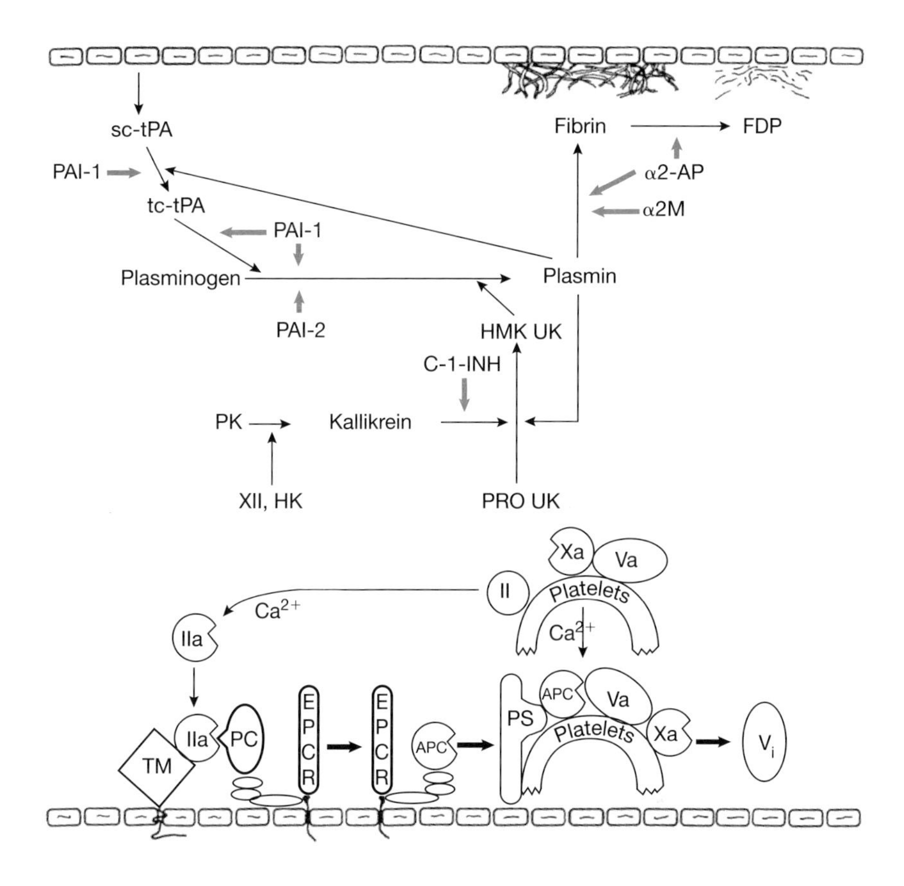


**Figure 4. Schematic representation of the anticoagulant and fibrinolytic pathways of blood coagulation**

The fibrinolytic system is shown at the top and the protein C anticoagulant system is shown below within the lumen of the blood vessel lined by endothelial cells. Thrombin (IIa) is generated by the assembly of the prothrombinase complex (thrombin, factor Xa and factor Va) on the platelet membrane in the presence of $Ca^{2+}$ which then binds to thrombomodulin (TM) on the endothelial cell membrane. Protein C (PC) binds to endothelial cell protein C receptor (EPCR) and is activated to activated protein C (APC) by the TM–IIa complex. Activated protein C dissociates from EPCR and binds to PS (protein S) forming a complex on the platelet membrane that inactivates factor Va to (to factor $V_i$) or factor VIIIa (to factor $VIII_i$), resulting in the shut down of the coagulation pathway. The fibrinolytic system (shown at the top of the figure) is under strict control by activators and inhibitors (blue arrows). Endothelial cells secrete plasminogen activators, e.g. sc-tPA (single chain tPA), that initiate the dissolution of the clot by converting clot-bound plasminogen into active plasmin. Plasmin degrades and dissolves the fibrin clot into fibrin degradation products (FDPs). The contact system also plays a role in the enhancement of fibrinolysis. α2-AP, α-2-antiplasmin; α2-M, α-2-macroglobulin; C-1-INH, C1 inhibitor; tc-tPA, two-chain t-PA; HMK-UK, high molecular mass urokinase or two-chain urinary-type plasminogen activator; PRO UK, pro-urokinase or single-chain urinary-type plasminogen activator.

generation, because the cofactors that are essential for increased catalytic efficiency of factor X and prothrombin activation are destroyed rapidly [3]. Important links between blood coagulation and inflammation, including the important role of the protein C anticoagulant system in the control of inflammation [23], have been emphasized recently. The clinical importance of the protein C anticoagulant system is demonstrated by the increased risk of thrombosis observed in patients with deficiencies of protein C and protein S, and by the mutation of the factor V molecule to a form that is not efficiently inactivated by activated protein C (i.e. factor V-Leiden, factor V(Arg$^{506}$→Gln).

**SERPINs, kunins and other inhibitors**

A variety of protease inhibitors are present in circulating blood [24–27]; these include SERPINs and kunins. The SERPINs (e.g. antithrombin III, α-1-protease inhibitor, heparin cofactor 2, α-2-antiplasmin, C1 inhibitor, ovalbumin) belong to a superfamily of proteins whose primary and tertiary structures are highly conserved. The mechanism of inhibition of serine proteases by SERPINs involves cleavage of the reactive centre of the inhibitor by the active site of the serine protease with a subsequent major change in structure that is characterized by loop insertion and the formation of a high-affinity complex with the enzyme [24–26]. The target proteases recognized by

**Table 2. Protease inhibitors in blood coagulation, anticoagulation and fibrinolysis**

*Estimated concentrations of PN-2 in plasma after secretion from platelets (3 × 10$^8$/ml).

| Protease inhibitors | Molecular mass (kDa) | Plasma concentration | | Target proteases |
|---|---|---|---|---|
| | | mg/l | nM | |
| Kunins | | | | |
| TFPI | 40 | 0.01 | 0.025 | FVIIa, FXa |
| PN-2 | 100–120 | <0.007 (0.2–2) | <0.06 (3–30)* | FXIa, FIXa |
| SERPINs | | | | |
| Antithrombin III | 60 | 290 | 470 | FIIa, FXa, FIXa |
| Heparin cofactor II | 66 | 40 | 60 | FIIa |
| α-1-Protease | 55 | 2500 | 4500 | FXIa, elastase |
| C1-inhibitor | 105 | 240 | 170 | FXIIa, FXIa, kallikrein, plasmin, C1s (complement system) |
| α-2-Antiplasmin | 70 | 70 | 105 | Plasmin |
| PAI-1 | 52 | 0.05 | 0.2 | t-PA, urokinase |
| PAI-2 | 48 | <0.005 | 0.1 | t-PA, urokinase |
| Protein C inhibitor | 57 | 5 | 90 | Protein C, kallikrein |
| Other | | | | |
| $\alpha_2$-Macroglobulin | 720 | 2500 | 3400 | Non-specific |

the various SERPINs are summarized in Table 2. The most relevant SERPIN for regulation of blood coagulation is antithrombin III, which has high-affinity heparin-binding sites that potentiate the activity of the SERPIN against its target proteases [24–26]. The major function of SERPINs appears to be the inhibition of circulating proteases, such as factor XIa, factor XIIa, kallikrein, plasmin, factor Xa, factor IXa and thrombin, so as to prevent disseminated intravascular coagulation.

Another important class of inhibitors are the kunins, members of the Kunitz family [27], which are characterized by the presence of Kunitz-type protease inhibitory domains (see Table 2). These consist of approx. 56 amino acids, with three pairs of highly conserved cysteine residues that form three intramolecular disulphide bridges that are responsible for the observed functional stability of kunins. One of the most important members of this family is TFPI, which circulates in plasma bound mainly to the lipoproteins, contains three Kunitz protease inhibitor domains, and rapidly, effectively and reversibly inhibits both factor VIIa and factor Xa in complex with tissue factor [16,27]. Another kunin is the Alzheimer's β-amyloid precursor protein, the soluble form of which is PN-2 [18,27]. PN-2 contains one Kunitz-type protease inhibitor domain that serves as a highly effective, tightly binding inhibitor of factor XIa ($K_i \approx 400$ pM) whose activity is potentiated 10-fold by the presence of heparin ($K_i \approx 30$ pM). In contrast with TFPI, the concentration of PN-2 in plasma is well below the $K_i$ for inhibition of factor XIa. The main source of PN-2 appears to be platelet α-granules, which secrete PN-2 in high concentrations in response to specific stimuli. This secreted PN-2 effectively inhibits solution phase factor XIa but not factor XIa when bound to the platelet surface [18,20]. Thus, it appears that PN-2 is an effective regulator of solution phase blood coagulation, while cell-mediated coagulation triggered by factor XIa is unaffected, so that subsequent coagulation reactions are localized to the haemostatic thrombus [18,20].

Another potentially physiologically relevant regulatory mechanism involves two recently characterized proteins [28], one of which is a γ-carboxyglutamic acid-containing vitamin-K-dependent protein, protein Z, and the other of which is a SERPIN, protein Z-dependent protease inhibitor (ZPI). It has been demonstrated recently that protein Z is a cofactor for factor XIa inhibition by ZPI, even though ZPI inhibits factor XIa directly without potentiation by protein Z [28]. Investigations to determine the potential physiological relevance of protein Z and ZPI in the pathogenesis of human thrombosis are currently in progress.

Finally, the α-macroglobulins are unique in their ability to inhibit proteases from various different mechanistic classes by an interesting and unique mechanism that involves a 'bait-and-trap' mechanism [27]. $\alpha_2$-Macroglobulin is present in human plasma at concentrations of 2–5 μM and may be involved in the clearance of a variety of different classes of plasma proteases.

## Fibrinolytic mechanism

Fibrinolysis, or the plasminogen–plasmin enzyme system, is an integral part of the finely tuned countervailing mechanisms in the haemostatic system [29]. The interactions of proteins involved in fibrinolysis are presented schematically in Figure 4. As is the case with the procoagulant and anticoagulant pathways, the fibrinolytic mechanism involves the assembly of serine proteases and cofactor molecules on surfaces and the action of counter-regulatory inhibitor molecules that keep the blood-clot-dissolving mechanism in check. The fibrinolytic system seems to have evolved to accomplish a localized dissolution of the haemostatic thrombus and to prevent excessive fibrin accumulation at a time after arrest of bleeding has been accomplished in order to promote wound healing by re-establishment of blood flow. The major components of the fibrinolytic mechanism are plasminogen, a single-chain glycoprotein with a molecular mass of 92000 Da, which can be proteolytically activated by two physiologically relevant activators, tissue plasminogen activator (t-PA) and urokinase-type plasminogen activator, to generate the enzyme plasmin. Plasmin then proteolytically cleaves fibrin, beginning with the C-terminal portion of its α-chain, producing new C-terminal lysine residues, thereby enhancing the binding of fibrin to plasminogen. Both t-PA and urokinase-type plasminogen activator are very poor activators of plasminogen in solution, but the catalytic efficiency of t-PA-catalysed plasminogen activation is several hundred times greater when the ternary complex of the enzyme and substrate is assembled on the fibrin surface. Control mechanisms exist for limiting excessive fibrinolysis, as with the blood coagulation and anticoagulant pathways (discussed above). These involve PAI-1 and PAI-2, α-2-antiplasmin and $\alpha_2$-macroglobulin [29] as well as plasma carboxypeptidase B, which is referred to as thrombin-activatable fibrinolysis inhibitor, which inhibits the activation of plasminogen by removing C-terminal lysine residues from fibrin that are essential for plasminogen binding to fibrin and for its conversion into active plasmin [30]. The thrombin required for activation of thrombin-activatable fibrinolysis inhibitor and downregulation of fibrinolysis requires the feedback activation of factor XI by the initially low levels of thrombin that are generated via the tissue factor pathway [31].

## Conclusion

The haemostatic mechanism consists of a complex series of interactions between enzymes, cofactors, receptor molecules and substrates on cell surfaces and results in the generation and dissolution of haemostatic thrombi by mechanisms that involve regulated blood coagulation, anticoagulation and fibrinolysis. A detailed understanding of the biochemistry and physiology of blood coagulation, anticoagulation and fibrinolysis, and of the cellular receptors that promote their assembly, will be required to elucidate the pathogenesis of thrombotic and haemorrhagic diseases and to develop safe and effective therapies for their treatment and prevention.

## Summary


- *The serine proteases involved in coagulation, anticoagulation and fibrinolysis have all evolved from a common trypsin-like ancestral protease and are characterized by the presence of highly conserved modular protein domains that have similar functions in different proteins.*
- *The cofactor and cell receptor molecules that participate in the assembly of enzymic complexes are also modular in design and highly conserved in structure and function.*
- *The series of reactions that comprise blood coagulation are considered in the sequence in which they are thought to occur physiologically, with the tissue factor pathway initiating the activation of factor X after vascular injury, which results in the generation of small quantities of thrombin with subsequent rapid downregulation by TFPI.*
- *The small quantities of thrombin that are generated then activate coagulation factor XI with subsequent sequential activation of factor IX, factor X and prothrombin, which leads to the generation of large quantities of thrombin and the conversion of fibrinogen into fibrin; all these reactions occur on the surface of activated platelets.*
- *The regulatory and anticoagulant pathways of blood coagulation include the protein C anticoagulant mechanism. This involves the association of thrombin with the endothelial cell surface receptor, thrombomodulin. This complex converts protein C, which is bound to the endothelial cell protein C receptor, into activated protein C, an anticoagulant protein that proteolytically inactivates both factor Va and factor VIIIa, thus downregulating blood coagulation.*
- *Regulatory mechanisms involving SERPINs and kunins in the plasma comprise important regulatory pathways that limit disseminated intravascular coagulation.*
- *The fibrinolytic mechanism or the plasminogen–plasmin enzyme system prevents excessive fibrin accumulation by promoting the localized dissolution of thrombi after haemostasis has occurred, thereby promoting wound healing by re-establishment of blood flow.*


Work reported in this essay was funded by research grants from the National Institutes of Health (HL56914, HL56153 and HL46213).

## References


1. Sadler, J.E. & Davie, E.W. (1994) Hemophilia A, hemophilia B, and von Willebrand disease. In *The Molecular Basis of Blood Diseases* (Stamatoyannopoulos, G., Nienhuis, A.W., Majerus, P.W. & Varmus, H., eds), pp. 657–700, W.B. Saunders Company, Philadelphia, PA
2. Arun, B. & Kessler, C.M. (2001) Clinical manifestations and therapy of the hemophilias. In *Hemostasis and Thrombosis. Basic Principles and Clinical Practice* (Colman, R.W., Hirsh, J., Marder, V.J., Clowes, A.W. & George, J.N., eds), pp. 815–824, Lippincott Williams & Wilkins, Philadelphia, PA
3. Dahlback, B. & Stenflo, J. (1994) The protein C anticoagulant system. In *The Molecular Basis of Blood Diseases* (Stamatoyannopoulos, G., Nienhuis, A.W., Majerus, P.W. & Varmus, H., eds), pp. 599–627, W.B. Saunders Company, Philadelphia, PA
4. Esmon, C.T. (2001) Protein C, protein S, and thrombomodulin. In *Hemostasis and Thrombosis. Basic Principles and Clinical Practice* (Colman, R.W., Hirsh, J., Marder, V.J., Clowes, A.W. & George, J.N., eds), pp. 335–353, Lippincott Williams & Wilkins, Philadelphia, PA
5. Walsh, P.N. (1974) Platelet coagulant activities and hemostasis: a hypothesis. *Blood* **43**, 597–605
6. Walsh, P.N. (1994) Platelet-coagulant protein interactions. In *Hemostasis and Thrombosis: Basic Principles and Clinical Practice* (Colman, R.W., Hirsh, J., Marder, V.J. & Salzman, E.W., eds), pp. 629–651, J.B. Lippincott Company, Philadelphia, PA
7. Tracy, P.B. (2001) Role of platelets and leukocytes in coagulation. In *Hemostasis and Thrombosis: Basic Principles and Clinical Practice* (Colman, R.W., Hirsh, J., Marder, V.J., Clowes, A.W. & George, J.N., eds), pp. 575–596, Lippincott Williams & Wilkins, Philadelphia, PA
8. Greenberg, D.L. & Davie, E.W. (2001) Blood coagulation factors: their complementary DNAs, genes, and expression, In *Hemostasis and Thrombosis. Basic Principles and Clinical Practice* (Colman, R.W., Hirsh, J., Marder, V.J., Clowes, A.W. & George, J.N., eds), pp. 21–57, Lippincott Williams & Wilkins, Philadelphia, PA
9. Patthy, L. (1985) Evolution of the proteases of blood coagulation and fibrinolysis by assembly from modules. *Cell* **41**, 657–663
10. Davie, E.W. & Ratnoff, O.D. (1964) Waterfall sequence for intrinsic blood clotting. *Science* **145**, 1310–1312
11. MacFarlane, R.G. (1964) An enzyme cascade in the blood clotting mechanism, and its function as a biochemical amplifier. *Nature (London)* **202**, 498–499
12. Gailani, D. & Broze, Jr, G.J., (1991) Factor XI activation in a revised model of blood coagulation. *Science* **253**, 909–912
13. Naito, K. & Fujikawa, K. (1991) Activation of human blood coagulation factor XI independent of factor XII. Factor XI is activated by thrombin and factor XIa in the presence of negatively charged surfaces. *J. Biol. Chem.* **266**, 7353–7358
14. Walsh, P.N. (2001) Factor XI. In *Hemostasis and Thrombosis: Basic Principles and Clinical Practice* (Colman, R.W., Hirsh, J., Marder, V.J., Clowes, A.W. & George, J.N., eds), pp. 191–202, Lippincott Williams & Wilkins, Philadelphia, PA
15. Colman, R.W. (2001) Contact activation pathway: inflammatory, fibrinolytic, anticoagulant, anti-adhesive and antiangiogenic activities. In *Hemostasis and Thrombosis: Basic Principles and Clinical Practice* (Colman, R.W., Hirsh, J., Marder, V.J., Clowes, A.W. & George, J.N., eds), pp. 103–122, Lippincott Williams & Wilkins, Philadelphia, PA
16. Morrissey, J.H. (2001) Tissue factor: an enzyme cofactor and a true receptor. *Thromb. Haemostasis.* **86**, 66–74
17. Giesen, P.L., Rauch, U., Bohrmann, B., Kling, D., Roque, M., Fallon, J.T., Badimon, J.J., Himber, J., Riederer, M.A. & Nemerson, Y. (1999) Blood-borne tissue factor: another view of thrombosis. *Proc. Natl. Acad. Sci. U.S.A.* **96**, 2311–2315
18. Walsh, P.N. (2001) Roles of platelets and factor XI in the initiation of blood coagulation by thrombin. *Thromb. Haemostasis.* **86**, 75–82

19. Baglia, F.A. & Walsh, P.N. (2000) Thrombin-mediated feedback activation of factor XI on the activated platelet surface is preferred over contact activation by factor XIIa or factor XIa. *J. Biol. Chem.* **275**, 20514–20519
20. Scandura, J.M., Zhang, Y., Van Nostrand, W.E. & Walsh, P.N. (1997) Progress curve analysis of the kinetics with which blood coagulation factor XIa is inhibited by protease nexin-2. *Biochemistry* **36**, 412–420
21. Hantgan, R.R., Simpson-Haidaris, P.J., Francis, C.W. & Marder, V.J. (2001) Fibrinogen structure and physiology. In *Hemostasis and Thrombosis. Basic Principles and Clinical Practice* (Colman, R.W., Hirsh, J., Marder, V.J., Clowes, A.W. & George, J.N., eds), pp. 203–232, Lippincott Williams & Wilkins, Philadelphia, PA
22. Loewy, A.G., McDonagh, J., Mikkola, H., Teller, D.C. & Yee, V.C. (2001) Structure and function of factor XIII. In *Hemostasis and Thrombosis. Basic Principles and Clinical Practice* (Colman, R.W., Hirsh, J., Marder, V.J., Clowes, A.W. & George, J.N., eds), pp. 233–247, Lippincott Williams & Wilkins, Philadelphia, PA
23. Esmon, C.T. (2001) Role of coagulation inhibitors in inflammation. *Thromb. Haemostasis.* **86**, 51–56
24. Bock, S.C. (2001) Antithrombin III and heparin cofactor II. In *Hemostasis and Thrombosis. Basic Principles and Clinical Practice* (Colman, R.W., Hirsh, J., Marder, V.J., Clowes, A.W. & George, J.N., eds), pp. 321–365, Lippincott Williams & Wilkins, Philadelphia, PA
25. Broze, G.J. & Tollefsen, D.M. (1994) Regulation of blood coagulation by protease inhibitors. In *The Molecular Basis of Blood Diseases* (Stamatoyannopoulos, G., Nienhuis, A.W., Majerus, P.W. & Varmus, H., eds), pp. 629–656, W.B. Saunders Company, Philadelphia, PA
26. Carrell, R.W. & Boswell, D.R. (1986) Serpins: the superfamily of plasma serine proteinase inhibitors. In *Proteinase Inhibitors* (Barrett, A.J. & Salvesen, G., eds), pp. 403–420, Elsevier, Amsterdam
27. Pizzo, S.V. & Wu, S.M. (2001) Alpha-macroglobulins and kunins. In *Hemostasis and Thrombosis. Basic Principles and Clinical Practice* (Colman, R.W., Hirsh, J., Marder, V.J., Clowes, A.W. & George, J.N., eds), pp. 367–379, Lippincott Williams & Wilkins, Philadelphia, PA
28. Broze, Jr, G.J., (2001) Protein Z-dependent regulation of coagulation. *Thromb. Haemostasis* **86**, 8–13
29. Bachmann, F. (2001) Plasminogen–plasmin enzyme system. In *Hemostasis and Thrombosis. Basic Principles and Clinical Practice* (Colman, R.W., Hirsh, J., Marder, V.J., Clowes, A.W. & George, J.N., eds), pp. 275–320, Lipppincott Williams & Wilkins, Philadelphia, PA
30. Bajzar, L., Manuel, R. & Nesheim, M.E. (1995) Purification and characterization of TAFI, a thrombin-activatable fibrinolysis inhibitor. *J. Biol. Chem.* **270**, 14477–14484
31. Bouma, B.N. & Meijers, J.C. (1999) Fibrinolysis and the contact system: a role for factor XI in the down-regulation of fibrinolysis. *Thromb. Haemostasis.* **82**, 243–250

# 9

# Anatomy and pathology of HIV-1 peptidase


## Ben M. Dunn[1]

*Department of Biochemistry and Molecular Biology, University of Florida College of Medicine, P.O. Box 100245, Gainesville, FL 32610-0245, U.S.A.*



### Abstract

The peptidase of the HIV type 1 (HIV PR) is required for the replication of and further infection by the virus. A concerted effort has taken place in the past 15 years to understand the properties of this enzyme, as it serves as an excellent drug target for control of the virus. Owing to drug pressure, many mutations arise during turnover of the virus and some of these lead to resistance to the effects of the inhibitors. Recent advances in the understanding of the changes these mutations cause to the enzyme and its interaction with substrates and inhibitors have been described. In addition, studies of closely related retroviral enzymes from simian immunodeficiency virus, feline immunodeficiency virus and HIV-2 have expanded the structure–function paradigm. The role of the flexibility of ligands and of the enzyme in active-site interactions is discussed.


### Introduction

In the past 15 years the human immunodeficiency virus/HIV has become a part of our lexicon. After the discovery of the virus in the 1980s, work began immediately to identify targets for antiviral drug therapy. The HIV peptidase (also known as HIV PR and HIV protease) was the second target to be identified, cloned and expressed. HIV peptidase is embedded within the


[1]*E-mail: bdunn@college.med.ufl.edu*

gag–pol fusion protein that is synthesized as a result of a frameshift in the mRNA. The peptidase is responsible for cleaving itself out of the polyprotein and for the cleavage of eight other junctions within gag–pol. The fragments produced assemble to form the virion particle [the matrix protein, the capsid protein (CA) and the nucleocapsid protein (NC)] or have other enzymic activities that are required for replication (reverse transcriptase and integrase). Thus, the initial proteolytic processing, which is catalysed by HIV peptidase, is absolutely essential to the life cycle of the virus and hence, is a viable candidate for drug therapy.

After its identification as a member of the aspartic peptidase family, strategies for drug discovery evolved rapidly, and focused initially on inhibitors of renin that had been developed previously by many pharmaceutical companies [1]. With suitable synthetic modification of those compounds followed by rigorous testing in culture, animals, and, eventually, humans, several molecules have survived to become clinically useful drugs. This has resulted in a dramatic improvement in the survival times of HIV-positive patients and a general improvement in the quality of life. Unfortunately, the emergence of viruses that are resistant to the first therapeutic drugs has led to a cycle of redesign and testing of new drugs in an effort to stay one step ahead of the virus [2]; this battle continues today.

The efforts that have been made to develop new therapeutics for treatment of HIV-positive patients has been a fascinating story in modern chemistry/biology, and several excellent reviews are available to describe the drug discovery process [1,2]. This chapter focuses on a separate issue involving HIV-1 peptidase; the fact that the enzyme has been studied by so many methods and with so many mutants means that we now have a very complete catalogue of the structure–function relationships of the properties of this protein. Of equal importance is the fact that HIV-1 peptidase has had its structure determined so many times that it might hold a world record in crystallography [3]. In-house groups in pharmaceutical companies have determined many of the structures and therefore these structures are not available publicly. Nevertheless, sufficient structures are available in the Research Collaboratory for Structural Bioinformatics Protein Database (and in [4]) to provide a complete view of the enzyme in various states. A third important factor has been the generous support by a variety of funding agencies (especially the National Institute of Allergy and Infectious Diseases of the National Institutes of Health) for research into the nature of inhibition of the enzyme.

This review focuses on the reports of changes in activity and binding properties of mutant forms of HIV-1 peptidase. It will also include, where appropriate, studies of closely related retroviral enzymes such as the peptidases from Rous sarcoma virus (RSV), equine infectious anaemic virus (EIAV), simian immunodeficiency virus (SIV), HIV-2 and feline immunodeficiency virus (FIV). Mutants that are created naturally by error-prone reverse transcriptase activity are selected by drug pressure that results in the outgrowth of resistant forms. Many studies have catalogued the points within HIV-1 peptidase where amino-acid changes lead to the drug resistance (summarized at http://hiv-web.lanl.gov/). Other studies have created point mutants through manipulation of DNA in the laboratory. Studies have created efficient expression vectors to produce reagent quantities of protein and purification schemes have been optimized to yield pure enzymes. A wide variety of experimental techniques have been applied to the study of these variants in addition to diffraction methods.

## Structure of HIV-1 peptidase and other related retroviral enzymes

The structure of HIV-1 peptidase can be described as a series of loops that fit together to form the 'structural template' of the retroviral enzymes [3]. These are presented in Figure 1 and are labelled for each of the identical monomers. Elements A1/A1* and A2/A2* are long hairpins that comprise elements that are external to the structure. Segments B1/B1* are critical, wide loops that make up the most central part of the structure. Loops B1 and B1* each have one aspartic residue that together form the catalytic machinery for this type of peptidase. Loops B1 and B1* come together at the middle of the molecule so that amino acids 25 and 25* are approx. 3 Å apart. Loops B2 and B2* provide important contact points for the binding of substrates and inhibitors. Loops C1 and C1* occupy positions that are more remote from the binding cleft. Elements C2 and C2* are helical portions. Segments D1 and D1* are long β-hairpins that form the flaps that overhang the active-site region. D1 and D1* also provide important interacting residues for contacts with substrates and inhibitors. Finally, D2 and D2* provide the four strands that form a β-sheet at the bottom of the molecule. In other retroviral enzymes, some of the loops can be larger or smaller than those found in HIV-1 peptidase. The C1 and C1* loops are helical segments in EIAV peptidase, for example.

In HIV peptidase, the amino acids that comprise these segments and the positions of significant mutations are shown in Table 1. Table 1 also shows are the positions of mutations that have an effect on substrate or inhibitor binding (illustrated in Figure 2). Some of these point mutations have little or no effect by themselves, but when combined with other point mutations they lead to a decrease in the binding of antiviral agents.

## Studies of mutant forms of HIV-1 peptidase

Swanstrom and co-workers [5] pioneered the approach of saturation mutagenesis of proteins, combined with a genetic selection system to screen rapidly for positions that are sensitive to substitution of natural amino acid residues. An example is given in [5], where this technique was applied to residues 46–56 which form the flaps (D1 and D1* in Figure 1). This region of each monomer contains a mixture of hydrophobic residues interspersed with glycine residues. The glycines contribute to a β-turn between two anti-parallel β-strands. As expected, positions with side chains that point outwards into the

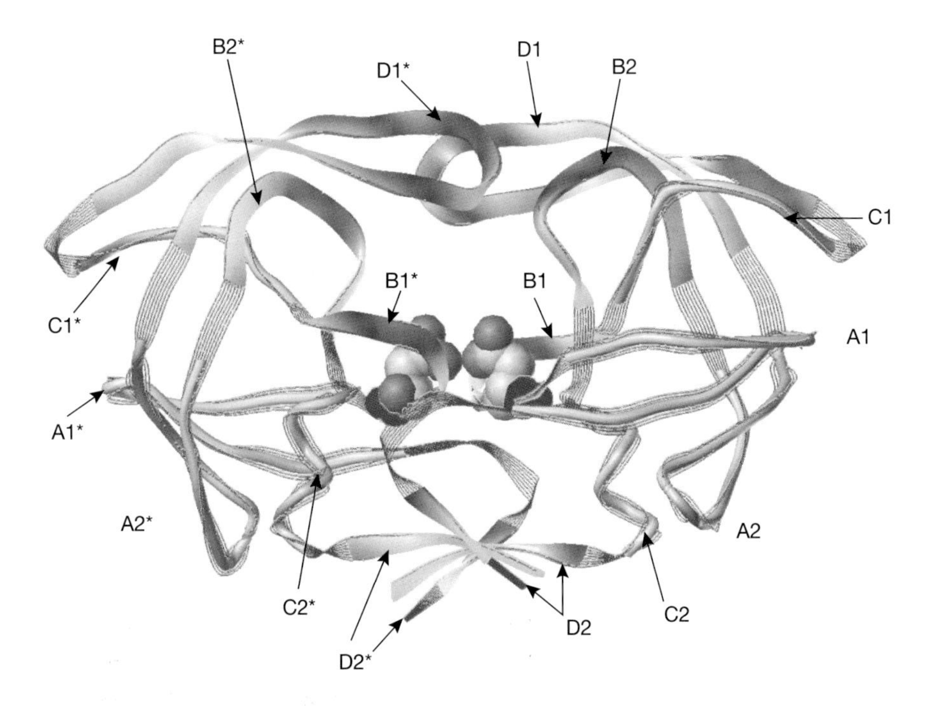


**Figure 1. Schematic representation of HIV-1 peptidase**
The protein consists of two monomers, arranged in this view so that one is on the right-hand side and one is on the left-hand side of the figure. An eight-strand line ribbon is drawn through the entire backbone of the enzyme. Individual structural elements are represented either as tubes or shaded ribbons, and alternate in each monomer. Each monomer is made up of four elements (A1–D1) that are repeated (A2–D2). The elements of the left-hand monomer are designated with asterisks to distinguish them from those of the right-hand monomer. The amino acids that comprise each element are given in Table 1. The catalytic aspartate residues are represented in space-filling fashion in the centre of the molecule.

**Table 1. Regions of sequence comprising structural segments in HIV-1 peptidase shown in Figure 1, and the positions of critical amino acid substitutions that affect substrate/drug binding**

| Segment (render) | Amino acid residues | Critical residues |
|---|---|---|
| – | – | 8, 10 |
| A1 (tube) | 10–22 | 20 |
| – | – | 23 |
| B1 (shaded ribbon) | 24–30 | 24, 27–30 |
| C1 (tube) | 32–38 | 32, 33, 36 |
| D1 (shaded ribbon) | 42–58 | 46–50, 54 |
| A2 (tube) | 61–73 | 63, 71, 73 |
| B2 (shaded ribbon) | 76–83 | 81, 82, 84 |
| C2 (tube) | 87–93 | 88 |
| – | – | 90, 91 |
| D2 (shaded ribbon) | 95–99+1–4 | – |

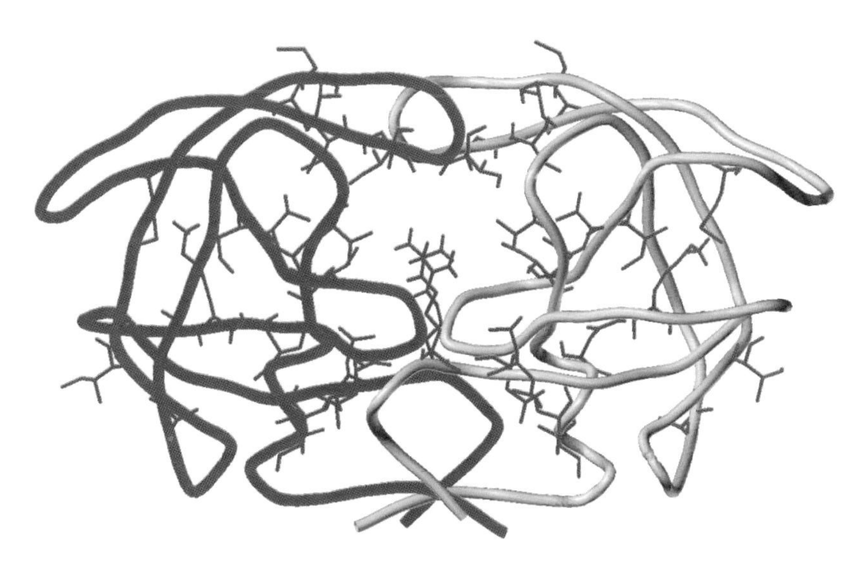

**Figure 2. HIV-1 peptidase and the positions of mutations that lead to drug resistance**
The positions of mutations that lead to drug resistance are indicated by capped-stick rendering. Note that while some substitutions are found within the active site cavity, many appear in regions remote from the binding cleft. Residues represented are 8, 10, 20, 24, 30, 32, 33, 36, 46, 47, 48, 50, 54, 63, 71, 73, 77, 82, 84, 90 and 91.

solvent were most tolerant of substitutions, while those with side chains that point inwards (towards the binding cavity) were least tolerant. However, even the outward-facing amino acids have some restrictions. The amino acids at positions 46 and 53 must be hydrophobic amino acids (Met and Phe respectively), as these interact to provide an important contribution to protein stability. $Gly^{49}$, $Gly^{51}$ and $Gly^{52}$ were intolerant of substitution, except that the $Gly^{51}$→Ala substitution gave an intermediate phenotype. $Gly^{48}$ accommodated several substitutions; with $Gly^{48}$→Glu yielding decreased processing of reverse transcriptase at the internal cleavage site. For the hydrophobic residues that point inward ($Ile^{47}$, $Ile^{50}$, $Ile^{54}$ and $Val^{56}$), several conservative substitutions were allowed, while substitutions that altered the hydrophobic nature were not allowed.

In addition to HIV-1, HIV-2 has been isolated from some patients; it appears to be less pathogenic than HIV-1. Sardana et al. [6] evaluated the differences between HIV-1 and HIV-2 peptidases, and found that only four amino acids (at positions 32, 47, 76 and 82) were changed. They created mutant forms of HIV-1 peptidase where the residue present in the HIV-2 enzyme was placed into the sequence. The $k_{cat}/K_m$ value for the $Val^{32}$→Ile mutant was 15-fold lower than the value for the wild-type enzyme. The $k_{cat}/K_m$ value for the $Ile^{47}$→Val mutant was decreased by 5-fold compared with the wild-type enzyme; the value for the $Leu^{76}$→Met mutant was approximately equal to that of the wild-type enzyme, while the $Val^{82}$→Ile mutant's $k_{cat}/K_m$ was twice that of wild-type. The binding of inhibitors was also affected most significantly by the Val$^{32}$→Ile substitution. This form of the enzyme is less active catalytically (see above), so that a mutant form of the virus with the Val$^{32}$→Ile alteration would not necessarily be viable. This study also demonstrated that compensatory mutations of the other three residues could restore the catalytic efficiency of the enzyme, while also improving the binding of the inhibitors.

Ridky et al. [7] studied a collection of HIV-1 peptidase point mutants in an effort to learn about the substrate preferences of these variant enzymes. The mutations Arg$^{8}$→Lys, Val$^{32}$→Ile, Val$^{82}$→Thr and Ile$^{84}$→Val had been observed in patients undergoing anti-retroviral therapy with peptidase inhibitors. The double substitutions Val$^{82}$→Thr/Ile$^{84}$→Val and Gly$^{48}$→Val/Leu$^{90}$→Met were also prepared. Using a substrate based on the CA–p2 (p2 is a small protein between CA and NC) cleavage site of the gag–pol polyprotein (Pro-Ala-Arg-Val-Leu↓Ala-Glu-Ala-Met-Arg, where ↓ indicates the cleavage site), the single point mutants were shown to be equivalent to the wild-type enzyme, with the exception of the Val$^{32}$→Ile and Arg$^{8}$→Lys variants, whose $k_{cat}/K_m$ values were 10-fold and 5-fold lower respectively. Substitutions were made in the RSV peptidase to alter residues at positions similar to those in the HIV peptidase. Interestingly, the inverse substitution to the Val$^{32}$→Ile mutation in HIV peptidase was constructed by making the Ile$^{44}$→Val mutation in RSV peptidase. The effects of decreasing or increasing the volume occupied by these residues caused changes in the selection of preferred substrates that were consistent with steric principles, but also that suggested that the influence can be felt in neighbouring subsites as well.

These authors also studied the effects of replacing the arginine residue in the P3 position in the substrate sequence (see Chapter 1, Figure 2, for an explanation of the terminology) with phenylalanine, and the effect of replacing the alanine residue at P1′ with methionine, leucine, phenylalanine, valine or isoleucine. In all cases, the substitutions in the substrate produced improvements in the $k_{cat}/K_m$ values, indicating that substrate selection remains the same in single-point or the two double-mutant forms of HIV-1 peptidase. The improved activity on a substrate with a larger hydrophobic amino acid at P1′ agrees with data obtained from the virus growing in the presence of inhibitors. Variant viruses with altered sequence in HIV-1 peptidase have been found with changes in the amino acids in certain cleavage sites, suggesting a compensation for the weakened catalytic activity of the variant peptidases.

In a follow-up study, Mahalingam et al. [8] evaluated several additional HIV-1 peptidase mutants for catalytic activity, structural stability and three-dimensional structure. In this work, the Arg$^{8}$→Gln, Asp$^{30}$→Asn, Gly$^{48}$→Val, Lys$^{45}$→Ile, Met$^{46}$→Leu, Val$^{82}$→Ser, Asn$^{88}$→Asp and Leu$^{90}$→Met mutants, when studied with several cleavage site peptides, showed variable catalytic activity. The Val$^{82}$→Ser mutant was the most affected, with catalytic activity between 2% and 20% that of the wild-type enzyme. The Arg$^{8}$→Gln, Asn$^{88}$→Asp and Leu$^{90}$→Met mutants had activities that were reduced by 60–80%, and the Gly$^{48}$→Val mutant was 50–80% as active as wild-type. The mutants Lys$^{45}$→Ile and Met$^{45}$→Leu showed a range of activities that exceeded that of wild-type enzyme for some substrates. In contrast, the Asp$^{30}$→Asn mutant exhibited a variable response depending on the substrate being analysed. Cleavage of a substrate based on the peptidase–reverse transcriptase junction was most affected, with activity being reduced by 10-fold; however, the Asp$^{30}$→Asn variant was fully active on some substrates. In terms of structural stability, as evaluated by the concentration of urea required to reduce activity by 50%, the Asp$^{30}$→Asn and Val$^{82}$→Ser variants were similar to wild-type, the Arg$^{8}$→Gln, Gly$^{48}$→Val, and Leu$^{90}$→Met variants showed decreased stability, and the stability of the Asn$^{88}$→Asp and Lys$^{45}$→Ile mutants increased, compared with the wild-type.

Structural determinations for three variants, Arg$^{8}$→Gln, Lys$^{45}$→Ile, and Leu$^{90}$→Met, were carried out. Comparison of these structures with the wild-type structure revealed that the number of hydrophobic intersubunit contacts provided the best correlation with the stability of the proteins, as measured by urea unfolding midpoints. The Lys$^{245}$→Ile mutant had the largest number of hydrophobic contacts between monomers and the Leu$^{90}$→Met variant had the least, which is directly related to stability. The ionic contact between Arg$^{8}$ and Asp$^{29}$ on the opposite subunit is lost in the mutant Arg$^{8}$→Gln, and is replaced by a water-mediated hydrogen bond.

## Energetics of inhibitor binding

An important energetic analysis of inhibitor binding has been provided by the work of Freire and co-workers [9]. In their study, calorimetric methods were used to separate the energy of binding into enthalpic and entropic components. The binding of indinavir, nelfinavir, saquinavir, and ritonavir, four clinically-approved drugs used in anti-retroviral therapy, was studied for both wild-type HIV peptidase and the Val$^{82}$→Phe/Ile$^{84}$→Val double mutant. The mutation positions are located on the B2 and B2* loops that provide significant contacts to substrates and inhibitors. The calorimetric studies found that, for all inhibitors except ritonavir, the enthalpy of binding to wild-type HIV-1 peptidase was unfavourable (positive). Ritonavir had a slightly favourable enthalpy of −9.6 kJ/mol (−2.3 kcal/mol). Therefore, to account for the nanomolar-to-subnanomolar $K_i$ values observed for these inhibitors, it is necessary to recognize that the entropy contribution to the overall free-energy change has to be very large and favourable (positive). This effect arises owing to the loss of water upon the transfer of the inhibitor molecules from aqueous solution to the hydrophobic cavity of the enzyme. For molecules that are flexible (i.e. that can rotate around single bonds) in solution, the binding process will include a unfavourable loss of entropy upon binding. However, for the peptidase-directed inhibitors that were studied in this work, the structural design included conformational restrictions on the molecular flexibility, which minimizes the unfavourable entropy component related to conformational mobility.

In comparing the binding process for the variant enzyme to the wild-type results, it was found that the enthalpic contribution became even more positive, while the entropic change did not deviate from the wild-type measurement. The $Val^{82}$→Phe and $Ile^{84}$→Val substitutions lead to a distortion of the binding cleft. Energy derived from van der Waals contacts is decreased, leading to fewer favourable enthalpic contributions to the overall enthalpy. The result is a less negative overall free energy of binding and a loss of binding.

## Structural studies of mutant enzymes

Prabu-Jeyabalan et al. [10] have provided a thorough analysis of substrate binding through the study of an inactive variant of HIV-1 peptidase ($Asp^{25}$→Asn) binding to a peptide based on the CA–p2 cleavage site in the gag–pol polyprotein, Lys-Ala-Arg-Val-Leu↓Ala-Glu-Ala-Met-Ser. This structure was refined to a crystallographic R-factor of 19.7% at a resolution of 2.0 Å. Significantly, only one orientation of the peptide chain was found in the structure. In many other structural determinations, ligands can bind in both orientations. For example, a structure of FIV peptidase, in which the active-site aspartate residue was changed to an asparagine, has been obtained by Laco et al. [11], but the electron density that was found indicated a disorder in the binding of the ligand. As the ligand in the case of HIV [10] was oriented in one direction, this allowed the authors to analyse the differences in binding at sets of identical sub-sites, P1/P1′, P2/P2′, P3/P3′ and P4/P4′. The ligand sequence is not symmetrical, so that different amino acids are bound at P1 (leucine) and P1′ (alanine), at P2 (valine) and P2′ (glutamate), at P3 (arginine) and P3′ (alanine), and at P4 (alanine) and P4′ (methionine).

To accommodate the different side chains of the ligand, the side chains of the enzyme adopt unique rotation angles. The distinction between the large, small, hydrophobic, or charged amino acids in different pockets provides an appreciation of how the enzyme might interact successfully with different amino acids at P2, for example for the nine different cleavage junctions found in the HIV gag–pol polypeptide. This adaptability is a unique feature of the retroviral peptidases.

Half of the side chains of the substrate point towards one side of the cleft and they consist of large and, in two cases, polar amino acids. The amino acids that point towards the other side of the cleft are small and hydrophobic. To provide additional insight into substrate binding, Prabu-Jeyabalan et al. [12] have studied the binding of five additional peptides that were designed based on the cleavage junctions within the gag–pol fusion protein, again using the inactive $Asp^{25}$→Asn mutant form of the enzyme. Comparison of the six complex structures revealed a common volume occupied by each sequence. This has provided clues to the design of a 'minimal' structure that should be effective against drug-resistant variants of HIV peptidase.

Analysis of the binding of substrate analogues to mutant forms of HIV-1 peptidase has been reported by Mahalingam et al. [13]. The point mutants Asp[30]→Asn, Lys[45]→Ile, Asn[88]→Asp and Leu[90]→Met were complexed with 'reduced peptide bond' analogues of the CA–p2 and the p2–NC cleavage sites, Arg-Val-Leu-r-Phe-Glu-Ala-Nle, and acetyl-Thr-Ile-Nle-r-Nle-Gln-Arg respectively, where -r- represents -$CH_2$-NH- in place of the peptide bond between the two amino acids that flank this symbol (Nle is norleucine). In processing of the gag–pol protein, the rate of cleavage at p2–NC has been shown to be approx. 400-fold faster than any other cleavage event, with the CA–p2 cleavage site having the lowest rate of processing.

All the complex structures show a similar binding geometry, with residues P2-P1-P1′-P2′ all superimposing well; root mean square deviation = 0.1–0.3 Å. The amino acids at P3 and P3′ show greater variability in the positions that they adopt. Of the mutants that were examined, two positions (Lys[45]→Ile and Asp[30]→Asn) make direct contacts with inhibitor molecules, while the other two (Asn[88]→Asp and Leu[90]→Met) do not. These complexes were compared with complexes of the same inhibitors with wild-type enzyme.

The substitution of isoleucine for lysine at position 45 in segment D1 leads to a more rigid flap structure. This leads to greater stability and higher activity against some substrates. Analysis of the binding mode of inhibitors revealed that isoleucine makes different contacts than the normal lysine side-chain, thereby increasing the hydrophobicity of the S4 binding site. The Asp[30]→Asn substitution in segment B1 leads to changes in interactions with hydrophobic side chains in the S2 positions of substrates and inhibitors, but polar interactions with Asp[30] are maintained in the Asn mutant. This could suggest that these interactions are vital to the function of the enzyme.

Leu[90] in segment C2 is near the dimer interface and replacement by methionine leads to steric problems owing to the longer side chain. The dimer interface includes interactions between the Asp[25] residues in each monomer at the catalytic site. Thus, replacing leucine with methionine leads to unfavourable interactions that can be correlated to the reduced stability of the mutant dimer. Asn[88] is also part of the C2 helix. In some complexes, Asn[88] is linked to Asp[30] through water molecules. Therefore, it is to be expected that patients undergoing treatment with nelfinavir develop mutations at both locations of HIV-1 peptidase, and such cases have been reported.

## Studies of FIV peptidase

After the determination of the three-dimensional structure of the FIV peptidase [14], there has been a great deal of interest in the similarities and differences between the FIV and HIV-1 peptidases. Lin et al. [15] prepared mutants of FIV peptidase in an effort to make its specificity closer to that of HIV peptidase. This was based on the comparison of the two active sites determined by X-ray diffraction methods, which showed that, while the overall geometries were identical in the binding cleft, there were several amino acids that differed between the two enzymes. In some cases, these differences were identical with mutations that arise in HIV-1 peptidase upon drug challenge. Thus, the FIV peptidase can be viewed as a model system for the drug-resistant forms of HIV-1 peptidase. This was confirmed by the observation that FIV peptidase does not bind to inhibitors of HIV-1 peptidase, and that it cleaves HIV-1 cleavage junction peptides weakly.

The mutations $Ile^{37}$→Val, $Asn^{55}$→Met, $Val^{59}$→Ile and $Gln^{99}$→Val each replaced the residue in FIV peptidase with the amino acid found at the same position in three-dimensional space in the HIV-1 enzyme. All these mutants had increased activity against HIV-1 cleavage sites, and bound HIV-1 peptidase-specific inhibitors better than wild-type FIV peptidase. However, other single replacements created enzymes with reduced or no activity. Thus, the initial assumption of a direct correspondence might be wrong, or interactions within the binding cleft are much more complicated than originally suspected.

Two efforts have been made to correlate binding among retroviral enzymes. Li et al. [16] studied complexes of both FIV and HIV peptidases with a C2-symmetrical compound that had (1*S*,2*R*,3*R*,4*S*)-1,4-diamino-1,4-dibenzyl-2,3-diol in the P1-P1′ position. This central feature was flanked by Val-Ala-Cbz on the N-terminal side and Cbz-Ala-Val on the C-terminal side (where Cbz is benzyloxycarbonyl). This molecule bound to FIV peptidase with a $K_i$ of $41 \pm 7$ nM and to HIV-1 peptidase with a $K_i$ of $1.5 \pm 0.3$ nM. This difference in binding to HIV-1 peptidase and FIV peptidase was also observed in studies of non-symmetrical inhibitors [17]. The mutants $Val^{59}$→Ile and $Gln^{99}$→Val were also studied using crystallography. Three important correlations have emerged. First, there are small, but significant, differences in the positions of the benzyl side chains in positions P1 and P1′ in the inhibitor when comparing the four complexes; secondly, the binding of the Cbz groups and adjacent peptide bonds at P4-P4 and P4′-P5′ is distinctly different in FIV peptidase compared with HIV-1 peptidase. Finally, two ‘structural’ water molecules can be found in all four complexes, but the exact positions are dependent upon the size of the P3 substituent. The authors suggest that design of new inhibitors could be influenced by these observations.

Kervinen et al. [18] have studied the binding of a different inhibitor, acetyl-naphthylalanine-Val-statine-aminobutyric acid naphthylalanine amide, to the HIV-1, FIV and EIAV peptidases. Val-P2 to 2-amino butyrate-P2′ interacts with the three enzymes in a similar fashion. Although the amino acids in the three enzymes in the S2–S2′ subsites are different, there is enough flexibility owing to side chain rotation to accommodate the inhibitor through this middle part. However, in distinct contrast, the naphthylalanine residues at the P3 and P3′ positions bind in very different ways to the three enzymes. Differences that can be as large as 120° in the side-chain rotation angles provide alternative binding modes in both S3 and S3′ subsites.

## Comparison of HIV-1 and SIV peptidases

Hoog et al. [19] studied the binding of an inhibitor, SB203386, to a triple mutant of HIV-1 peptidase as well as to SIV peptidase. Differences in $K_i$ values of up to 6-fold could be accounted for by small changes in active site binding. However, an additional 9-fold factor is required to account for the total difference in measured $K_i$ values. These authors postulated that shifts in surface loops were responsible, specifically in the region of residue 34 (Figure 3). Towler et al. [20], following up on the report of Hoog et al., investigated the amino-acid residues that confer specificity in binding of inhibitors. In addition to active site residues 32, 47, and 82, they found that amino acids in the region of residues 31–37, which lie outside the active site cavity, had an effect on events within the active site. Swairjo et al. [21] carried this analysis further by crystallizing HIV-1 peptidase chimaeras in which the residues from HIV peptidase at 31 and 33–37 were substituted. The key substitution was the replacement of $Glu^{35}$ in the HIV-1 sequence with glycine. This increases the flexibility of the loop region C1, as well as destroying a salt bridge between $Glu^{35}$ and $Arg^{57}$. $Arg^{57}$ is at the start of the D1 hairpin loop that interacts with the active site. Thus, a substitution in a surface loop can have consequences that are translated into the size and shape of the active-site cavity. This correlates nicely with observations of mutations in many positions outside the active site in forms of HIV-1 peptidase that arise during drug therapy.

**Figure 3. HIV-1 peptidase and the 31–37 loop**
In the left-hand monomer, the backbone is depicted as a white ribbon, while the 31–37 loop is shown in blue. In the monomer on the right, the colours are reversed.

## Whole molecule as a unit

Rose et al. [22] have proposed a global hypothesis to account for the development of resistance in HIV-1 peptidase. Their analysis compared the structures of unliganded and liganded HIV-1 and SIV enzymes. By studying all parts of the enzyme, they identified five domains (see Figure 4). The first of these is a terminal domain that consists of the N- and C-terminal segments (D2 and D2*; see Figure 1) and half of the helical segment (C2 and C2*). Two 'core' domains are made up of the catalytic region (loops B1) in addition to the 'P1' loop (loop B2) and some surface loops (A1 and A2). One core domain would consist of loops A1, A2, B1, and B2, and the second core domain would consist of loops A1*, A2*, B1* and B2*, i.e. each monomer contributes one core domain. Finally, loops D1 and D1* provide the two flap domains. Rotations of 4–7° between the terminal domain and the core domains have been observed in SIV and HIV-1 peptidases, while rotations of 4–6° are observed for movement of the flap domains relative to the core domains. Among other effects, these rotations cause significant changes in the width of the substrate-binding cleft.

In order to account for the development of drug resistance, Rose and colleagues have proposed that the binding and release of substrates and inhibitors take place at different rates. If mutations at the interfaces between domains lead to changes in the rates of movement of the domains and if these movements are important for ligand binding, then it might be expected that both types of

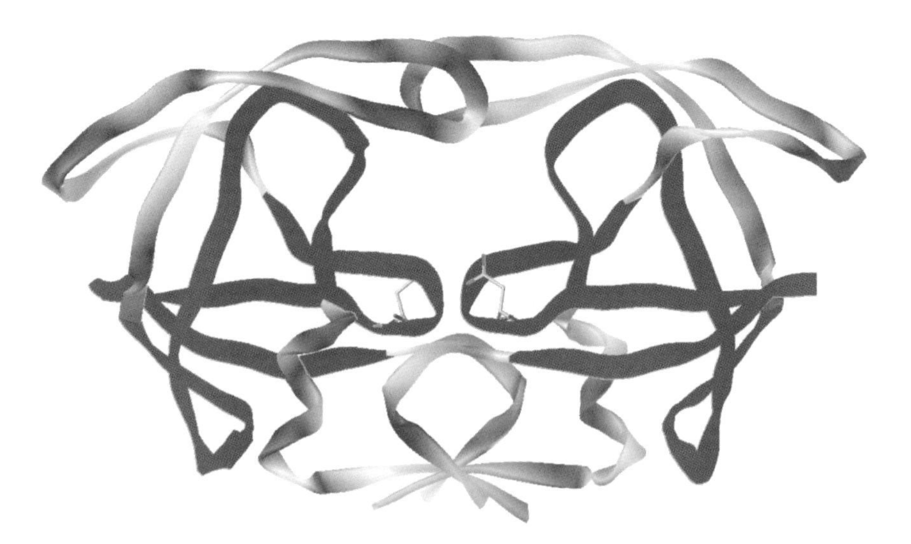

**Figure 4. HIV-1 peptidase and domain flexibility [21]**
The terminal domain is represented as a white ribbon (bottom), the central domain including the catalytic site is represented as a blue ribbon, and the flap domains are represented as white ribbons (top).

ligands would be affected in a similar way. However, these authors point out that substrates have an inherent flexibility that is often not shared by inhibitors that are developed against the HIV-1 peptidase. Substrate binding may have an advantage owing to this flexibility, as we have seen that the active site shows considerable adaptability (sometimes known as 'plasticity'). A flexible substrate molecule, particularly in the form of a polyprotein segment, may bind and undergo rearrangements leading to productive complexes, which leads to cleavage events. For a rigid inhibitor, adapting to the peptidase binding cleft may not proceed as smoothly, allowing the opportunity to dissociate before achieving the final, 'closed-and-locked' structure seen in X-ray studies. The provocative hypothesis provided by Rose and colleagues [22] should be testable by measurements of binding and dissociation rates of inhibitors and substrates. At the least, this proposal should stimulate new work in this important area.

## Summary

- *The peptidase of HIV is essential to viral replication and is therefore a viable drug target.*
- *Many mutant forms of the enzyme arise as a consequence of drug pressure.*
- *These mutant forms are resistant to the effects of therapeutic inhibitors, thus creating the need for understanding this phenomenon in order to design better inhibitors.*
- *A large body of work on structural aspects of the HIV peptidase and related retroviral enzymes, alone and in complex with inhibitors, has provided a broader view of protein structure–function.*
- *Characterization of the kinetic, thermodynamic, and structural properties of variant forms of HIV peptidase has provided important data on the highly malleable properties of this fascinating enzyme.*
- *Studies of FIV and SIV peptidases and comparisons to HIV peptidases have expanded the structure–function algorithm.*
- *Considering movements of structural units of the enzyme reveals global movements of the protein that can account for the effects of mutations remote from the active site.*

Work in the authors' lab on HIV-1 peptidase is funded by the National Institutes of Health grant AI-28571.

## References

1. Tomasselli, A.G. & Heinrikson, R.L. (2000) Targeting the HIV-protease in AIDS therapy: a current clinical perspective. *Biochim. Biophys. Acta* **1477**, 189–214
2. Swanstrom, R. & Erona, J. (2000) Human immunodeficiency virus type-1 protease inhibitors: therapeutic successes and failures, suppression and resistance. *Pharmacol. Ther.* **86**, 145–170

3. Wlodawer, A. & Gustchina, A. (2000) Structural and biochemical studies of retroviral proteases. *Biochim. Biophys. Acta* **1477**, 16–34
4. Vondrasek, J. van Buskirk, C.P. & Wlodawer, A. (1997) Database of three-dimensional structures of HIV proteinases. *Nat. Struct. Biol.* **4**, 8
5. Shao, W., Everitt, L., Manchester, M., Loeb, D.D., Hutchison, III, C.A. & Swanstrom, R. (1997) Sequence requirements of the HIV-1 protease flap region determined by saturation mutagenesis and kinetic analysis of flap mutants. *Proc. Natl. Acad. Sci.* U.S.A. **94**, 2243–2248
6. Sardana, V.V., Schlabach, A.J., Graham, P., Bush, B.L., Condra, J.H., Culbertson, J.C., Gotlib, L., Graham, D.J., Kohl, N.E., LaFemina, R.L. et al. (1994) Human immundeficiency virus type 1 protease inhibitors: evaluation of resistance engendered by amino acid substitutions in the enzyme's substrate binding site. *Biochemistry* **33**, 2004–2010
7. Ridky, T.W., Kikonyoho, A., Leis, J., Gulnik, S., Copeland, T., Erickson, J., Wlodawer, A. Kurinov, I., Harrison, R.W. & Weber, I. (1998) Drug-resistant HIV-1 proteases identify enzyme residues important for substrate selection and catalytic rate. *Biochemistry* **37**, 13835–13845
8. Mahalingam, B. Louis, J.M., Reed, C.C., Adomat, J.M. Krouse, J., Wang, Y.-F., Harrison, R.W. & Weber, I.T. (1999) Structure and kinetic analysis of drug resistant mutants of HIV-1 protease. *Eur. J. Biochem.* **263**, 238–245
9. Todd, M.J., Luque, I., Valazquez-Campoy, A. & Freire, E. (2000) Thermodynamic basis of resistance to HIV-1 protease inhibition: calorimetric analysis of the V82F/I84V active site resistant mutant. *Biochemistry* **39**, 11876–11883
10. Prabu-Jeyabalan, M. Nalivaika, E. & Schiffer, C.A. (2000) How does a symmetric dimer recognize an asymmetric substrate? A substrate complex of HIV-1 protease. *J. Mol. Biol.* **301**, 1207–1220
11. Laco, G.S., Schalk-Hihi, C., Lubkowski, J., Morris, G., Zdanov, A., Olson, A., Elder, J.H., Wlodawer, A. & Gustchina, A. (1997) Crystal structures of the inactive D30N mutant of feline immunodeficiency virus protease complexed with a substrate and an inhibitor. *Biochemistry* **36**, 10696–10708
12. Prabu-Jeyabalan, M., Nalivaika, E. & Schiffer, C.A. (2002) Substrate shape determines specificity of recognition for HIV-1 protease: analysis of crystal structures of six substrate complexes. *Structure* **10**, 369–381
13. Mahalingam, B., Louis, J.M., Hung, J., Harrison, R.W. & Weber, I.T. (2001) Structural implications of drug-resistant mutants of HIV-1 protease: high-resolution crystal structures of the mutant protease/substrate analogue complexes. *Proteins: Struct. Funct. Genet.* **43**, 455–464
14. Wlodawer, A., Gustchina, A., Reshetnikova, L., Lubkowski, J., Zdanov, A., Hui, K.Y., Angleton, E.L., Farmerie, W.G., Goodenow, M.M., Bhatt, D. et al. (1995) Structure of an inhibitor complex of the proteinase from feline immunodeficiency virus. *Nat. Struct. Biol.* **2**, 480–488
15. Lin, Y.-C., Beck, Z., Lee, T., Le, V.-D., Morris, G.M., Olson, A.J., Wong, C.-H. & Elder, J.H. (2000) Alteration of substrate and inhibitor specificity of feline immunodeficiency virus protease. *J. Virol.* **74**, 4710–4720
16. Li, M., Morris, G.M., Lee, T., Laco, G.S., Wong, C.-H., Olson, A.J., Elder, J.H., Wlodawer, A. & Gustchina, A. (2000) Structural studies of FIV and HIV-1 proteases complexed with an efficient inhibitor of FIV protease. *Proteins: Struct. Funct. Genet.* **38**, 29–40
17. Dunn, B.M., Pennington, M.W., Frase, D.C. & Nash, K. (1999) Comparison of inhibitor binding to feline and human immunodeficiency virus proteases: structure-based drug design and the resistance problem. *Biopolymers (Peptide Sci.)* **51**, 69–77
18. Kervinen, J., Lubkowski, J. Zdanov. A., Bhatt, D., Dunn, B.M., Hui, K.Y., Powell, D.J., Kay, J., Wlodawer, A. & Gustchina, A. (1998) Toward a universal inhibitor of retroviral proteases: comparative analysis of the interactions of LP-130 complexed with proteases from HIV-1, FIV, and EIAV. *Protein Sci.* **7**, 2314–2323
19. Hoog, S.S., Towler, E.M., Zhao, B., Doyle, M.L., Debouck, C. & Abdel-Meguid, S.S. (1996) Human immunodeficiency virus protease ligand specificity conferred by residues outside of the active site cavity. *Biochemistry* **35**, 10279–10286

20. Towler, E.M., Thompson, S.K., Tomaszek, T. & Debouck, C. (1997) Identification of a loop outside the active site cavity of the human immunodeficiency virus proteases which confers inhibitor specificity. *Biochemistry* **36**, 5128–5133
21. Swairjo, M.A., Towler, E.M., Debouch, C. & Abdel-Meguid, S.S. (1988) Structural role of the 30′s loop in determining the ligand specificity of human immunodeficiency virus protease. *Biochemistry* **37**, 10928–10936
22. Rose, R.B., Craik, C.S. & Stroud, R.M. (1998) Domain flexibility in retroviral proteases: structural implications for drug resistant mutations. *Biochemistry* **37**, 2607–2621

# 10

# Inhibition of peptidases in the control of blood pressure

Eiji Kubota, Rachel G. Dean, Leanne C. Balding and Louise M. Burrell

*Department of Medicine, University of Melbourne, Austin and Repatriation Medical Centre, Studley Road, Heidelberg 3084, Victoria, Australia*

## Abstract

The natriuretic peptide and renin–angiotensin systems are physiological counterparts with opposite roles in the regulation of electrolyte balance and blood pressure. In both systems, membrane-bound, zinc-dependent peptidases play an important role in the inactivation or activation of the system. Angiotensin-converting enzyme (ACE) converts angiotensin I into angiotensin II, and neutral endopeptidase (NEP) degrades the natriuretic peptides. Simultaneous inhibition NEP and ACE by a single molecule (a vasopeptidase inhibitor) is a new therapeutic approach in hypertension. Wider applications for vasopeptidase inhibitors being studied include their role as cardioprotective agents in heart failure, as renoprotective agents in chronic renal failure and diabetic nephropathy, and as vasculoprotective agents in endothelial dysfunction and athersclerosis.

## Introduction

The natriuretic peptide and renin–angiotensin systems are physiological counterparts with opposite roles in the regulation of electrolyte balance and

[1]*To whom correspondence should be addressed (e-mail: burrell@austin.unimelb.edu.au)*

blood pressure (Table 1) [1]. Natriuretic peptides are a family of peptides with potent physiological actions, which are released from the heart, brain and endothelial cells. The physiological effects include salt and water excretion (natriuresis and diuresis), relaxation of blood vessels, and antiproliferative and antihypertrophic activities [2]. In many ways, the natriuretic peptides can be regarded as endogenous inhibitors of the renin–angiotensin system as they oppose the actions of angiotensin II (Ang II) on vascular resistance, blood pressure and renal sodium homoeostasis, as well as blunting the secretion of aldosterone by Ang II. One action, however, that the natriuretic peptides and Ang II do share is that they are potent inhibitors of renin release from the kidney.

In both systems, membrane-bound, zinc-dependent peptidases play an important role in the inactivation or activation of the system; the ectoenzyme angiotensin-converting enzyme (ACE; EC 3.4.15.1) is responsible for the conversion of Ang I to Ang II [3], while neutral endopeptidase (NEP; EC 3.4.24.11) is a component of one of the pathways involved in the degradation of the natriuretic peptides [4].

Hypertension or high blood pressure represents a continuing challenge for a number of reasons:

- it is very prevalent and produces a significant global health burden;
- it affects 1 in every 6 members of the population and this incidence rises to 1 in every 2 in those aged 70 years and older;
- it is a major risk factor for heart disease, stroke and renal failure;
- control of blood pressure and continuation of therapy is less than optimal;
- new antihypertensive agents that provide better blood-pressure control reverse structural and functional abnormalities and that have improved tolerability may change this situation.

The use of ACE inhibitors has been a significant advance in the treatment of hypertension [5], while specific NEP inhibitors were developed several years ago [6]. Although these inhibitors were shown to elevate plasma levels of the natriuretic peptides and to cause the expected responses such as diuresis, natriuresis and peripheral vasodilatation under certain experimental conditions, clinical trials in hypertension and cardiac failure had disappointing results. This

**Table 1. Biological actions of Ang II and natriuretic peptides**

↑ indicates an increased activity; ↓ indicates decreased activity.

| Action | Ang II | Natriuretic peptides |
|---|---|---|
| Vasoconstriction | ↑ | ↓ |
| Sympathetic activity | ↑ | ↓ |
| Proliferation | ↑ | ↓ |
| Hypertrophy | ↑ | ↓ |
| Aldosterone release | ↑ | ↓ |
| Sodium excretion | ↓ | ↑ |
| Renin release | ↓ | ↓ |

was partly owing to the fact that a decrease in blood pressure activates the renin–angiotensin system, and any increase in natriuretic peptide levels, be it from infusion of natriuretic peptides or inhibition of their breakdown, is unable to overcome an activated renin–angiotensin system. However, in the presence of an inhibited renin–angiotensin system, the biological actions of natriuretic peptide are restored, and this led to the development of compounds that simultaneously inhibit NEP and ACE. Such compounds are known as vasopeptidase inhibitors, and may offer advantages in treating hypertension [7]. The addition of NEP to ACE inhibition potentiates the vasodilator, natriuretic and diuretic actions of natriuretic peptides.

## Renin–angiotensin system

One of the major advances in the field of the renin–angiotensin system has been the appreciation that it functions as a dual hormonal system, serving both as a circulating and a local-tissue hormone system. All components of the renin–angiotensin system are present in important cardiovascular structures, including the heart, vessels, brain, kidney and adrenal gland. The primary active hormone of the renin–angiotensin system, Ang II, is produced as a result of an enzymic cascade (Figure 1). The glycoprotein angiotensinogen is synthesized by the liver and it is cleaved by the enzyme renin to form Ang I within the circulation. Renin is secreted from the juxtaglomerular cells of the kidneys in response to decreases in blood volume, blood pressure or sodium concentration, and it is the rate-limiting enzyme in the final production of Ang II. In the last step, Ang I is cleaved by the metallopeptidase ACE to form the octapeptide Ang II in both the circulation and tissues. All the known effects of the renin–angiotensin system can be accounted for by the multiple actions of Ang II, which interacts with at least two known membrane receptors, the

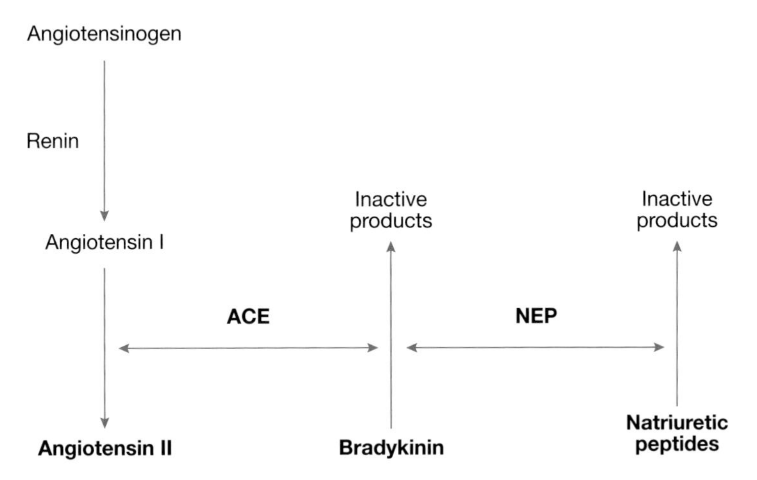


**Figure 1. The renin–angiotensin system cascade, and its interaction with kinins and the natriuretic peptide system**

angiotensin type 1 receptor and the angiotensin type 2 receptor. The well known physiological effects of Ang II such as vasoconstriction, aldosterone stimulation, and salt and water homoeostasis, appear to be mediated via stimulation of the G-protein-coupled angiotensin type 1 receptor.

# ACE

ACE is an ectoenzyme that anchors itself to the plasma membrane by its C-terminal end and has a large extracellular domain. ACE has two active catalytic sites, the N- and C-domains, which are both zinc-dependent but differ in their chloride requirements and catalytic constants. ACE is an integral membrane component of all endothelial cells, occurs in many epithelial cells (gastrointestinal tract, kidney, reproductive tract, placenta, brain), and is expressed on fibroblasts and macrophages [8]. ACE also has a wide substrate-specificity; ACE acts on bradykinin, substance P, opioid peptides, neurotensin, cholecystokinin, bombesin, enkephalins and luteinizing-hormone-releasing hormone *in vitro*, but the effect of ACE on such substrates *in vivo* is not fully understood.

Whether any of the reported benefits of ACE inhibitors involve bradykinin or other ACE substrates is still a highly controversial issue. Only comparative trials with the recently developed new class of Ang II receptor antagonists are likely to settle the issue. However, the data obtained thus far indicate that Ang II receptor antagonists, which do not interfere with bradykinin metabolism, have the same effect as the ACE inhibitors on morbidity and mortality in patients with hypertension and renal disease.

## ACE inhibitors and hypertension

ACE inhibitor is the generic name applied to a group of drugs that were first described in 1977 by Ondetti and Cushman [3] and that act by binding competitively to the active catalytic sites of ACE, thereby preventing access to the endogenous substrate. ACE inhibition depends on the co-ordination of the active site with ligands that are present on the inhibitor. All ACE inhibitors (Figure 2) act in an identical manner, but differ in the degree and the length of time of ACE inhibition in different tissues. To improve oral absorption, many second generation ACE inhibitors are ester prodrugs that are metabolized in the liver and gut-wall to release more active diacid derivatives; for example, enalapril is activated to produce the active component enalaprilat.

The indications for ACE inhibitors have widened considerably in scope since they were first introduced for the treatment of hypertension. A series of landmark studies (reviewed in [5]), which started in 1987 with the CONSENSUS study, showed that ACE inhibitors should be used in all grades of heart failure that are associated with left ventricular dysfunction. There is now evidence that ACE inhibitors reduce the rate of progression of renal failure in both diabetic and non-diabetic renal disease, effects that appear to go beyond

| Class | Structure |
|---|---|
| ***ACE inhibitors*** Captopril | 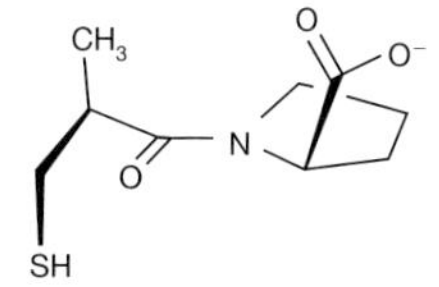 |
| Enalaprilat | 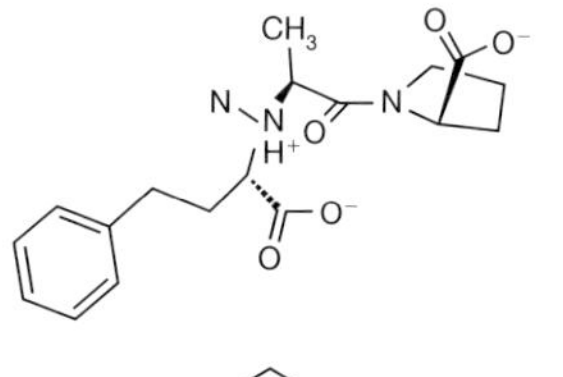 |
| ***NEP inhibitors*** Thiorphan | 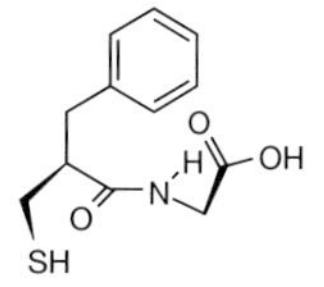 |
| SCH 424345 | 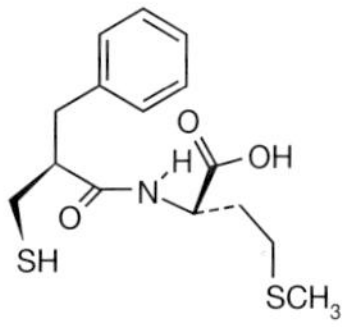 |
| ***NEP/ACE inhibitors*** Omapatrilat | 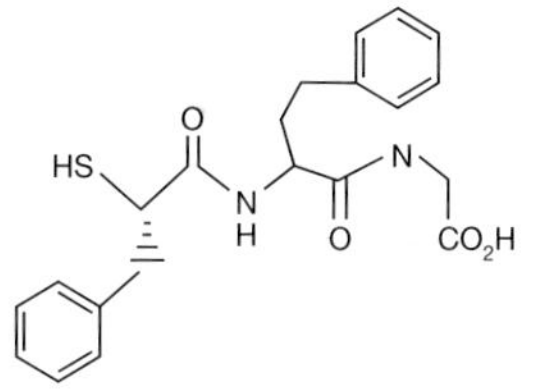 |
| MDL 100,173 | 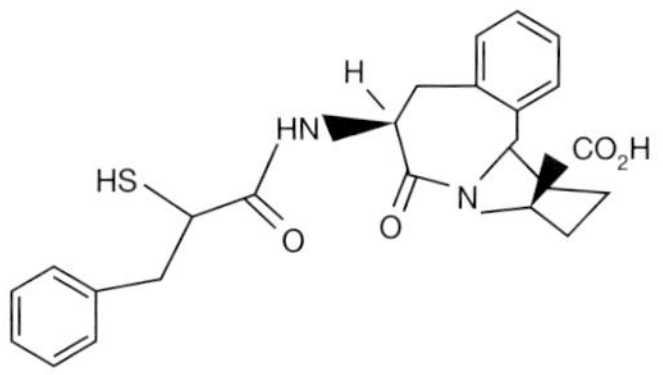 |

**Figure 2. Chemical structures of ACE, NEP and NEP/ACE inhibitors**

their effects on blood pressure. Ongoing studies will determine whether the mortality benefits seen in heart failure transfer to hypertension, and will clarify the potential effects of ACE inhibition on recurrent myocardial infarction and atherosclerosis.

Clinical trials have shown that antihypertensive treatment improves cardiovascular morbidity and mortality, but it is not known if the benefits are due solely to blood pressure reduction or are specific for different classes of drug. ACE inhibitors lower mean systolic and diastolic blood pressure in hypertensive patients irrespective of age, but they are less effective in hypertensive African–American patients compared with Caucasian patients. This may be related to low renin levels in African–Americans, and if ACE inhibitors are combined with a diuretic, which increases renin levels, they are as effective as in other ethnic groups.

The precise mechanism by which ACE inhibition causes a decrease in blood pressure remains unclear; the acute hypotensive effects of ACE inhibition relate with pre-treatment levels of renin and plasma Ang, but their long-term antihypertensive effect results from more complex mechanisms [9]. ACE has multiple substrates and interacts with other systems that are involved in cardiovascular regulation. Possible mechanisms of action include inhibition of ACE in, for example, the heart, kidneys, adrenal glands and blood vessels, a decrease in plasma aldosterone, suppression of the sympathetic nervous system and accumulation of bradykinin.

## Natriuretic peptides

The natriuretic peptides are a family of at least three structurally similar peptides including atrial natriuretic peptide (ANP), brain natriuretic peptide and C-type natriuretic peptide [2] (Figure 3). The human ANP gene, which was first sequenced in 1984, is located on chromosome 1 and consists of three exons separated by two introns. Human ANP is derived from a 151-amino-acid precursor, preproANP. Within cardiac myocytes, preproANP is processed rapidly into proANP, which has a high degree of identity between species. The human version of proANP (126 amino acids) is the major constituent of atrial secretory granules, which fuse with the cell surface to release their contents when exposed to hormone-release stimuli. During this process, proANP is thought to be cleaved, by an unidentified enzyme, to yield two products (Figure 3); the N-terminal end becomes $ANP_{1–98}$, and the 28 amino acids of the C-terminus become the biologically active $ANP_{99–126}$ (or ANP).

The physiological actions of ANP are mediated through guanylate-cyclase-linked receptors upon ligand binding. The natriuretic peptides act as endogenous antagonists of the renin–angiotensin system (Table 1) to cause natriuresis and diuresis, vasodilation and suppression of the sympathetic nervous system; they also inhibit cell growth and decrease the secretion of aldosterone and renin. Natriuretic peptides play an important role in the regulation of cardiovascular, renal and endocrine function, but the therapeutic potential for ANP in hypertension and heart failure is limited as ANP is a peptide and therefore, if administered orally, it will be degraded rapidly in the stomach before it can exert any actions. An alternative approach is to increase endogenous ANP levels by inhibition of its enzymic degradation by NEP.

## NEP

NEP, which was originally referred to as enkephalinase because of its ability to degrade opioid peptides within the brain, was subsequently shown to be

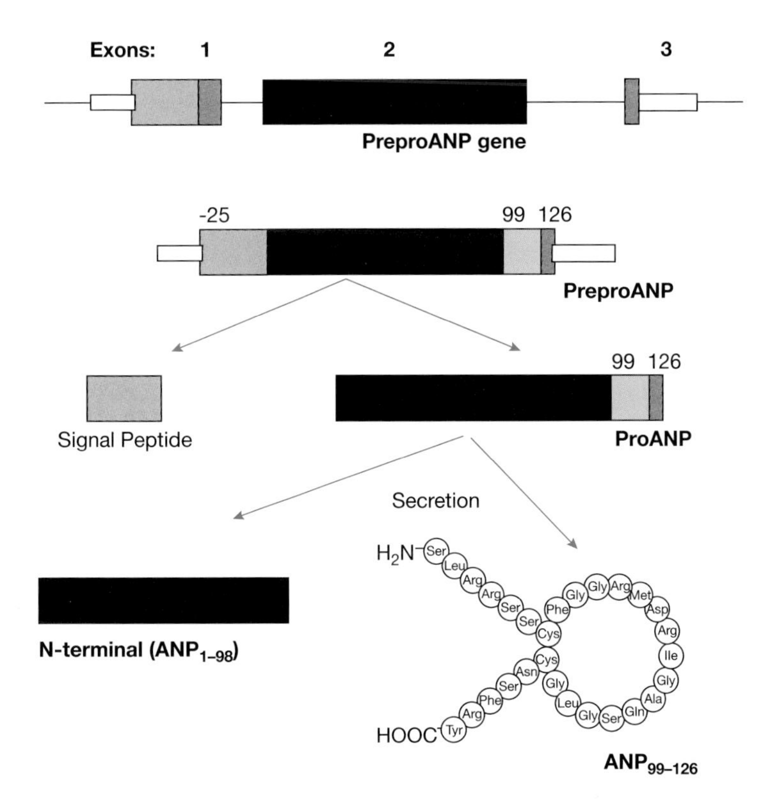


**Figure 3. Structure and expression of the ANP gene, and the processing pathway and release of ANP**

identical with an already well characterized zinc metallopeptidase that is present in the kidneys [10]. NEP is an integral membrane dimeric glycoprotein with a short intracellular domain, a transmembrane domain that anchors NEP in the plasma membrane and a large ectodomain that includes the zinc-containing active site. As a zinc-dependent metallopeptidase, NEP shares mechanistic similarities with other metallopeptidases including ACE, endothelin-converting enzyme, aminopeptidases and carboxypeptidases [6].

NEP hydrolyses peptide bonds on the amino side of hydrophobic amino-acid residues. In the case of ANP, NEP cleaves the $Cys^{105}$–$Phe^{106}$ bond to disrupt the ring structure and inactivate the peptide [11]. While the role of NEP in the inactivation of ANP has been most extensively studied, it has also been demonstrated to play a role in the metabolism of brain natriuretic peptide and C-type natriuretic peptide. NEP has broad substrate selectivity *in vitro* and it cleaves enkephalins, endothelin, substance P, kinins, neurotensin, the insulin B-chain, Ang II, calcitonin gene-related peptide and adrenomedullin, as well as the natriuretic peptides. One of the main functions of NEP *in vivo* is to metabolize the natriuretic peptides.

NEP is located principally within the kidney where it rapidly degrades filtered ANP, thereby preventing the peptide from reaching more distal luminal sites. NEP is also found in the lung, gut, liver, adrenal glands, brain, heart and vasculature, and is present on endothelial cells and on the surface of neutrophils and leukaemic cells.

### NEP inhibitors

NEP inhibitor (Figure 2) is the generic term applied to a group of drugs that act by binding competitively to the active site of NEP, thereby preventing access to the endogenous substrate. NEP inhibitors were developed as analgesics (pain killers) because the enzyme metabolizes enkephalins and increases opioid levels, but the current interest focuses on the role of NEP in the metabolism of the natriuretic peptides [6]. As with ACE inhibitors, inhibition of NEP depends on the co-ordination of the active site with ligands that are components of the inhibitor. Many clinically useful NEP inhibitors are prodrugs that are activated by hepatic esterases.

### NEP inhibition in hypertension

The natriuretic peptides are the major mediators of the cardiovascular and renal effects of NEP inhibition. NEP inhibitors have little effect on blood pressure, natriuresis or diuresis when ANP levels are normal, but when plasma ANP levels are elevated, NEP inhibition produces the expected physiological changes. Unfortunately, long-term NEP inhibition has minimal antihypertensive effects in patients with hypertension [12]; however, experimental studies in both hypertension and heart failure have shown that the addition of an ACE inhibitor restores the haemodynamic and renal benefits of NEP inhibition.

Compounds that simultaneously inhibit ACE and NEP have now been designed, taking into account the similar structural characteristics of the catalytic site of both enzymes (Figure 2). This novel class of drugs, which inhibit the renin–angiotensin system and potentiate the effects of the natriuretic peptides, are known as NEP/ACE inhibitors or vasopeptidase inhibitors [7,13]. Several of these inhibitors are available, including omapatrilat, candoxatril and MDL 100,173.

All of these compounds inhibit ACE and NEP, although the degree of potency against the individual enzymes varies between compounds (Figure 4). We have used radioligand binding assays with the specific NEP inhibitor radioligand [$^{125}$I]RB104 and the specific ACE inhibitor radioligand [$^{125}$I]MK351A to show that S21402 is a stronger inhibitor of renal NEP than it is of lung ACE *in vitro*. After oral dosing in rats, it was also found that S21402 was a more potent inhibitor of renal NEP than of renal ACE [14]. By contrast, the inhibition of NEP in rat renal tissue by the vasopeptidase inhibitor omapatrilat is similar to its inhibition of ACE in the same tissue [15]. Because the long-term benefits of ACE inhibitors in preventing or reversing target-organ damage depends on their ability to inhibit ACE in target organs such as the kidney and heart, it will be important to assess whether differences in the degree or the site of NEP inhibition are of importance in determining clinical efficacy.

## Vasopeptidase inhibitors and hypertension

Hypertension represents a continuing challenge because of the reasons listed in the introduction, and although ACE inhibitors are firmly established in the treatment of hypertension, 50% of patients require additional therapy. In experimental hypertension, the major advantage of the vasopeptidase inhibitors is their ability to decrease blood pressure independently of body volume or renin status [16], whereas ACE inhibitors are most effective in renin-dependent hypertension, and NEP inhibitors are most effective in low-renin, volume-dependent hypertension. Thus, it is expected that vasopeptidase inhibitors will be effective as single-drug therapies in a wider range of patients compared with selective ACE inhibition [7]. Indeed, early case reports that describe the use of omapatrilat have shown powerful, dose-dependent decreases in systolic and diastolic blood pressure, regardless of age, race or gender. Before such drugs do enter the clinical arena, large-scale randomized controlled trials in humans are necessary.

The precise mechanism by which vasopeptidase inhibition lowers blood pressure is not known. As for the selective ACE inhibitors, the long-term anti-hypertensive effect is likely to result from complex mechanisms, particularly given the multiple substrates for ACE and NEP. Possible mechanisms of action include not only decreasing plasma Ang II levels, but also the accumulation of natriuretic peptides and the inhibition of ACE and NEP at the tissue level. Autoradiographic studies allow the precise localization of NEP and ACE in tissues such as the heart, kidney and blood vessels, and can be used to assess inhibition of NEP and ACE after treatment with vasopeptidase inhibitors. Such studies are important in helping to determine the mechanisms of action of the vasopeptidase inhibitors, which are not clear at present.

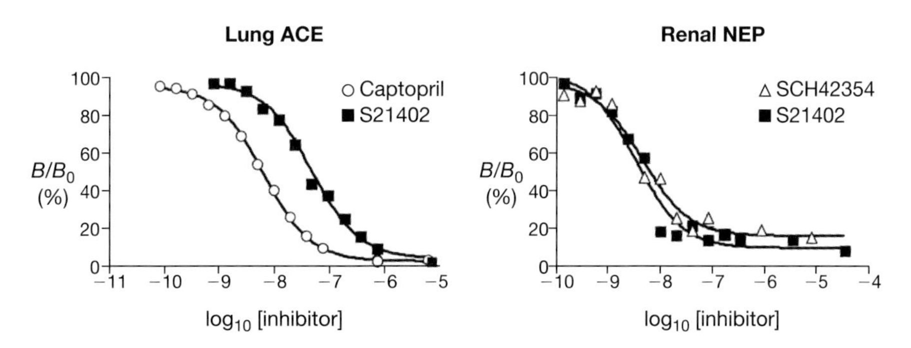


**Figure 4. Effects of a vasopeptidase inhibitor, S21402, compared with those of the selective ACE inhibitor captopril and the selective NEP inhibitor SCH42354 on the binding of the ligands [$^{125}$I]MK351A to lung ACE and [$^{125}$I]RB104 to renal NEP respectively**

*B*, observed binding; $B_0$, initial binding. Points indicate mean values; $n = 3$ per curve. The *x*-axes show the $\log_{10}$ of the concentration (mol/l) of the inhibitor.

**Table 2. Cardiovascular benefits of vasopeptidase inhibitors**

+, data available; − no data available.

| Benefit | Experimental models | Man |
|---|---|---|
| Decrease blood pressure | + | + |
| Decrease cardiac hypertrophy/fibrosis | + | − |
| Improve myocardial ischaemia | + | − |
| Regress atherosclerosis | + | − |
| Improve endothelial dysfunction | + | − |
| Improve mortality in heart failure | + | − |
| Slow progression of renal disease | + | − |

Hypertension is also an independent risk factor for renal failure, which is a major public health problem in the Western world. Recent guidelines from the World Health Organization/International Society of Hypertension emphasize that blood pressure levels need to be lower in patients with renal disease and diabetes than in other hypertensive groups [17]. Results from our laboratory suggest that the vasopeptidase inhibitors may help to achieve these more aggressive blood pressure targets, at least in animal models of renal disease [18]. Table 2 summarizes the current state of knowledge with regard to the use of the vasopeptidase inhibitors in experimental cardiovascular disease.

## Summary


- *The natriuretic peptide and renin–angiotensin systems are physiological counterparts.*
- *In both systems, membrane-bound, zinc-dependent peptidases play important roles in the inactivation or activation of the system.*
- *ACE converts Ang I into Ang II, and NEP degrades the natriuretic peptides.*
- *Simultaneous inhibition of the peptidases NEP and ACE by a single molecule (vasopeptidase inhibitor) is a new therapeutic approach in hypertension.*
- *Wider applications for vasopeptidase inhibitors include their role as cardioprotective agents in heart failure, as renoprotective agents in chronic renal failure and diabetic nephropathy, as well as vasculoprotective agents in endothelial dysfunction and atherosclerosis.*


## References


1. Johnston, C.I., Hodsman, P.G., Kohzuki, M., Casley, D.J., Fabris, B. & Phillips, P.A. (1989) Interaction between atrial natriuretic peptide and the renin angiotensin aldosterone system. Endogenous antagonists. *Am. J. Med.* **87**, 24S–28S
2. Levin, E.R., Gardner, D.G. & Samson, W.K. (1998) Mechanisms of disease – natriuretic peptides. *N. Engl. J. Med.* **339**, 321–328

3. Cushman, D.W. & Ondetti, M.A. (1999) Design of angiotensin converting enzyme inhibitors. *Nat. Med.* **5**, 1110–1112
4. Turner, A.J., Isaac, R.E. & Coates, D. (2001) The neprilysin (NEP) family of zinc metalloendopeptidases: genomics and function. *Bioessays* **23**, 261–269
5. Burrell, L.M. & Johnston, C.I. (2000) Clinical trials of ACE inhibitors: pathogenetic insights and therapeutic advances. In *Drugs, Enzymes and Receptors of the Renin–Angiotensin System: Celebrating a Century of Discovery* (Husain, A. and Graham, R.M., eds), pp. 191–200, Harwood, Amsterdam
6. Roques, B.P., Noble, F., Dauge, V., Fournie-Zaluski, M.-C. & Beaumont, A. (1993) Neutral endopeptidase 24.11: structure, inhibition and experimental and clinical pharmacology. *Pharmacol. Rev.* **45**, 118–146
7. Bralet, J. & Schwartz, J.-C. (2001) Vasopeptidase inhibitors: an emerging class of cardiovascular drugs. *Trends Pharmacol. Sci.* **22**, 106–109
8. Burrell, L.M., Chai, S.Y. & Johnston, C.I. (1996) Tissue and cellular renin–angiotensin systems In *Handbook of Hypertension* (Zanchetti, A. & Mancia, G., eds), pp. 822–841, Elsevier, Amsterdam
9. Johnston, C.I. (1994) Tissue angiotensin converting enzyme in cardiac and vascular hypertrophy, repair and remodeling. *Hypertension* **23**, 258–268
10. Kerr, M.A. & Kenny, A.J. (1974) The purification and specificity of a neutral endopeptidase from rabbit kidney brush border. *Biochem. J.* **137**, 477–488
11. Watanabe, Y., Nakajima, K., Shimamori, Y. & Fujimoto, Y. (1997) Comparison of the hydrolysis of the three types of natriuretic peptides by human kidney neutral endopeptidase 24.11. *Biochem. Mol. Med.* **61**, 47–51
12. Richards, A.M., Crozier, I.G., Kosoglou, T., Rallings, M., Espiner, E.A., Nicholls, M.G., Yandle, T.G., Ikram, H. & Frampton, C. (1993) Endopeptidase-24.11 inhibition by SCH-42495 in essential hypertension. *Hypertension* **22**, 119–126
13. Corti, R., Burnett, J.C., Rouleau, J.L., Ruschitzka, F. & Luscher, T.F. (2001) Vasopeptidase inhibitors – a new therapeutic concept in cardiovascular disease? *Circulation* **104**, 1856–1862
14. Farina, N.K., Johnston, C.I. & Burrell, L.M. (2000) Reversal of cardiac hypertrophy and fibrosis by S21402, a dual inhibitor of neutral endopeptidase and angiotensin converting enzyme in SHRs. *J. Hypertens.* **18**, 749–755
15. Burrell, L.M., Droogh, J., Manintveld, O., Rockell, M.D., Farina, N.K. & Johnston, C.I. (2000) Antihypertensive and antihypertrophic effects of omapatrilat in SHR. *Am. J. Hypertens.* **13**, 1110–1116
16. Trippodo, N.C., Robl, J.A., Asaad, M.M., Fox, M., Panchal, B.C. & Schaeffer, T.R. (1998) Effects of omapatrilat in low, normal and high renin experimental hypertension. *Am. J. Hypertens.* **11**, 363–372
17. Cooper, M.E. & Johnston, C.I. (2000) Optimizing treatment of hypertension in patients with diabetes. *JAMA, J. Am. Med. Assoc.* **283**, 3177–3179
18. Cao, Z.M., Burrell, L.M., Tikkanen, I., Bonnet, F., Cooper, M.E. & Gilbert, R.E. (2001) Vasopeptidase inhibition attenuates the progression of renal injury in subtotal nephrectomized rats. *Kidney Int.* **60**, 715–721

# 11

# Shedding of membrane proteins by ADAM family proteases

Marcia L. Moss*[1] and Millard H. Lambert†

*Cognosci Inc., 2 Davis Drive, P.O. Box 12076, Research Triangle Park, NC 27709, U.S.A., and †Department of Computational, Analytical and Structural Sciences, GlaxoSmithKline Research and Development, 5 Moore Drive, Research Triangle Park, NC 27709, U.S.A.

## Abstract

Many membrane-bound proteins undergo proteolytic release from the membrane, a process known as 'shedding'. Some of the processing events are carried out by enzymes of the ADAM (a disintegrin and metalloproteinase) family, which are also membrane bound. One of the most well known ADAM family members is TACE (tumour necrosis factor-α-converting enzyme. TACE was the first ADAM family member to have a known physiological substrate, namely, precursor tumour necrosis factor-α. Inhibitors of TACE block the release of the soluble form of this inflammatory cytokine, and are currently being studied in drug discovery projects for the treatment of arthritis. Since the discovery of TACE, physiological substrates for other ADAMs have been determined. This review focuses on the shedding events carried out by TACE and other ADAM family proteinases.

## Introduction

Many membrane proteins have soluble extracellular domains that are subject to proteolytic release, a process known as 'shedding'. Membrane proteins that

[1]*To whom correspondence should be addressed (e-mail: moss0610@yahoo.com).*

are subject to shedding typically have an extracellular domain anchored to the membrane by a single transmembrane segment. In Type I membrane proteins, such as the amyloid precursor protein (APP), the transmembrane segment lies near the C-terminus of the protein chain, whereas in Type II membrane proteins, such as precursor tumour necrosis factor α (TNF-α), the transmembrane segment lies near the N-terminal end of the protein chain [1]. In either case, shedding is effected by proteolytic cleavage in a 'stalk' region between the transmembrane segment and the globular extracellular domain. In principle, the cleavages associated with shedding could be carried out by any protease on the external side of the membrane. However, some proteases are tethered to the membrane, thereby effectively localizing their catalytic activity to the vicinity of susceptible membrane proteins. This review will focus specifically on shedding by the largest such class of membrane-tethered proteases, the ADAM (a disintegrin and metalloproteinase domain) family of zinc metalloproteases.

TNF-α provides a good example of ADAM-mediated shedding. Precursor TNF-α has a short N-terminal cytosolic segment, a single transmembrane segment, and a large C-terminal extracellular domain (Figure 1). The soluble extracellular domain is released by cleavage in the stalk segment between $Ala^{76}$ and $Val^{77}$. The membrane-bound, precursor TNF-α mediates responses by cell-to-cell contact, whereas the soluble, 'mature' TNF-α can be carried away from the cell to mediate a response at distant sites of action. Both forms are active and it is unclear as to what contribution each makes to disease states such as arthritis and cancer. However, the soluble form of TNF-α is involved in rheumatoid arthritis (RA), and is a principle mediator of sepsis and HIV cachexia [2]. The cleavage of precursor TNF-α is the key step in regulating the balance between the membrane-bound and soluble forms of TNF-α. The

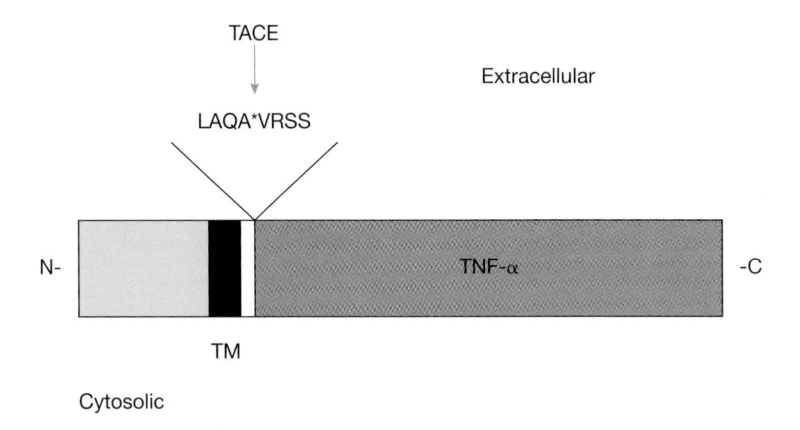


**Figure 1. Precursor TNF-α**
Processing of precursor TNF-α occurs near the transmembrane segment between positions $Ala^{76}$ and $Val^{77}$ by TACE to generate soluble TNF-α. TM, transmembrane region.

enzyme responsible for TNF-α shedding, the TNF-α-converting enzyme (TACE), has been cloned and expressed [3,4], and identified as a member of the ADAM family (ADAM 17).

## The ADAM family

The domain structure of the ADAM family of proteins, showing the disintegrin and metalloproteinase domains [5], is shown in Figure 2. Full-length ADAMs are Type I membrane proteins in which the catalytic domain is localized on the extracellular side of the membrane. Most of the ADAMs have a characteristic HEXXHXXGXXH zinc-binding motif in the metalloproteinase domain, which classifies the ADAM family within the 'metzincin' superfamily [6]. This superfamily also includes matrix metalloproteases (MMPs) (see Chapter 3) and the more closely related snake venom metalloproteases (SVMPs). In those ADAMs where the full HEXXHXXGXXH motif is present, the three histidine residues ligate a catalytic zinc atom, and the glutamic acid residue promotes catalysis by positioning and activating a water molecule to attack the peptide backbone [6]. Several of the ADAM family members have mutated zinc-binding motifs, but show good sequence identity elsewhere within the disintegrin and metalloproteinase domains [5]. These ADAMs are not expected to function as proteases, but could achieve biological effects through their other domains. The disintegrin domain may be involved in adhesion, as has been shown for some SVMPs (for a review see [7]), and may serve to recognize substrates in catalytically active ADAMs. ADAM family members have a pro domain at the N-terminal end of the protein chain. The corresponding pro domain in SVMPs

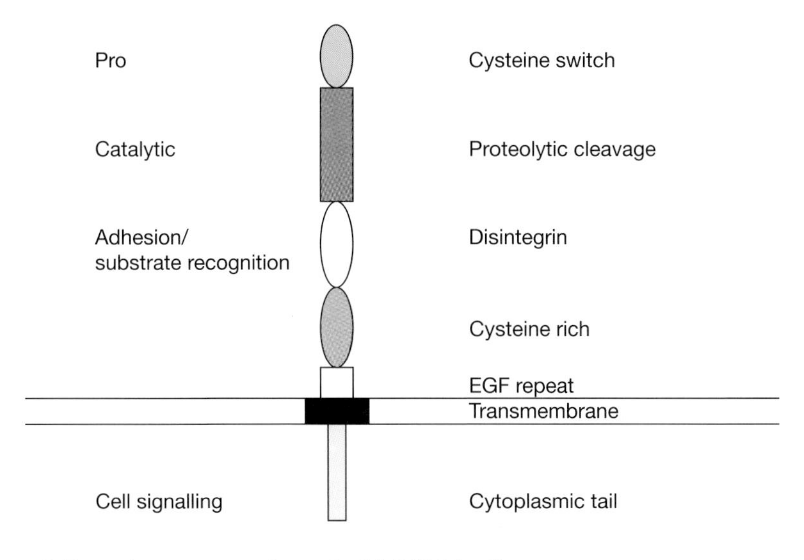


**Figure 2. Domain structure of a typical ADAM family protease**
The domain names are listed on the right-hand side, while their functions are listed on the left.

contains a cysteine residue that ligates the catalytic zinc atom, thereby maintaining the metalloproteinase domain in an inactive state by a 'cysteine switch' mechanism [8]. Most of the catalytically active ADAM family members have a similar cysteine-containing sequence in their pro domain, which suggests that their activity is also regulated by a cysteine-switch mechanism. Most of the ADAMs have a transmembrane segment and cytoplasmic tail at the C-terminal end of the chain. The cytoplasmic tail is subject to phosphorylation under certain conditions [9,10], can bind to SH3 (Src homology 3) domains (reviewed in [2,11]), and may therefore be involved in cell signalling. A list of ADAMs is maintained by Judith White at the University of Virginia (www.people.virginia.edu/~jag6n/Table_of_the_ADAMs.html); more than 25 different ADAMs have been identified in mammals to date.

## ADAM 17 (TACE)

As discussed above, ADAM 17, which is often referred to as TACE, has been identified as the protease responsible for shedding of TNF-α. TACE was originally purified based on its ability to process precursor TNF-α [3,4]. Mouse cells with an inactivated TACE gene ($TACE^{\Delta Zn/\Delta Zn}$) showed a marked reduction in soluble TNF-α production, confirming its role *in vivo* [3]. The TACE knockout work has also suggested a broader role for TACE in membrane protein shedding. For example, transforming growth factor α (TGF-α) release is impaired in fibroblasts that have been immortalized by ras transformation from $TACE^{\Delta Zn/\Delta Zn}$ mice [12]. In addition, shedding of other cell surface molecules, such as L-selectin, the TNF receptors I and II, the interleukin-6 receptor and APP, is affected in the TACE knockout animals or cells (reviewed in [2,11]). The $TACE^{\Delta Zn/\Delta Zn}$ knockout involves a deletion in the zinc-binding metalloprotease domain [3]. This deletion would remove the proteolytic activity of TACE, but could preserve its ability to interact with other proteins via the disintegrin domain, transmembrane segment or cytosolic tail. Thus, the $tace^{\Delta Zn}$ gene product could possibly bind to and sequester other proteins that would normally be processed by TACE or other ADAMs, thereby effectively functioning as a 'dominant negative' with respect to other ADAM family proteases. This effect complicates the interpretation of knockout studies. The heterozygous $TACE^{\Delta Zn/+}$ knockout can help to recognize dominant negative effects [12]. In addition, peptides that correspond to the hypothesized cleavage sites can be tested directly for cleavage by TACE. The available peptide data are summarized in Table 1. For example, while knockout studies have implicated TACE in the shedding of L-selectin [12] and APP [13], recombinant TACE cleaves the corresponding peptides relatively slowly, compared with a precursor TNF-α peptide (Table 1). This suggests that the $tace^{\Delta Zn}$ gene product is exerting a dominant negative effect, or that efficient cleavage by TACE requires protein–protein interactions not present in the peptide cleavage experiment.

**Table 1. Effect of recombinant TACE on peptide substrates**

An asterisk between the amino acids indicates where cleavage of the peptide substrate occurs. The specificity constant, $k_{cat}/K_m$, indicates how efficiently a particular substrate is cleaved by the enzyme; larger $k_{cat}/K_m$ values correspond to more efficient cleavage. DNP, dinitrophenyl; Ac, acetyl; ND, not determined

| Protein substrate | Peptide substrate | Cleavage (yes or no) | Specificity constant ($k_{cat}/K_m$) | Reference |
|---|---|---|---|---|
| Precursor TNF-α | DNP-SPLAQA*VRSSSR-$NH_2$ | Yes | $2 \times 10^5\ M^{-1} \cdot s^{-1}$ | [4] |
| TGF-α (C-terminus) | Ac-ADLLA*VVAAS-$NH_2$ | Yes | Poor substrate | [11] |
| | DNP-HADLLA*VVAASQ-$NH_2$ | Yes | $7 \times 10^3\ M^{-1} \cdot s^{-1}$ | |
| APP | Ac-VHHQK*LVFFA-$NH_2$ | Yes | Very poor substrate | [12] |
| TNF receptor I | DNP-LPQIEN*VKGTED-$NH_2$ | No | | |
| TNF receptor II | DNP-SMAPGA*VHLPQP-$NH_2$ | Yes | ND | Unpublished work |
| TRANCE | DNP-IVGPQR*FSGAPA-$NH_2$ | Yes | $2 \times 10^2\ M^{-1} \cdot s^{-1}$ | [2] |
| Notch | Ac-CPYKIEA*VKSEPV | Yes | ND | |
| Interleukin-6 receptor | DNP-TSLPVQ*DSSSVP-$NH_2$ | No | | |
| L-selectin | Ac-KLDK*SFSKIKEGDYN-$NH_2$ | Yes | Very poor substrate | [11] |

The use of multiple 'sheddases' with one of them being an ADAM family member, is not without precedent. APP is subject to cleavage at three general positions, α, β, and γ, which correspond to cleavage by an α-secretase, a β-secretase and a γ-secretase (reviewed in [14]; see Chapter 4). The γ-secretase-mediated cleavage occurs in the membrane, while the β cleavage occurs on the extracellular side of the membrane. Cleavage at the β- and γ-positions generates the β-amyloid peptide, which accumulates in the amyloid plaques of Alzheimer's disease. The α-secretase cleaves near the middle of the β-amyloid sequence, leading to a non-amyloidogenic soluble APPα (sAPPα) peptide. The β-secretase has been identified as a transmembrane aspartyl protease named BACE (β-site APP-cleaving enzyme), while the γ-secretase may be presenilin or a complex containing presenilin. The α-secretase has a constitutive activity as well as an activity that is regulated by protein kinase C. The knockout work described above suggested that TACE is a protein kinase C-regulated α-secretase. Studies with ADAM 10 (discussed below) suggest that it is also an α-secretase.

The structure of the catalytic domain of TACE has been determined by X-ray crystallography [15] (Figure 3a). In addition, several closely-related SVMP structures have been determined, including adamalysin II [16] (Figure 3B) and atrolysin C. The structures of TACE, adamalysin II and atrolysin C reveal a

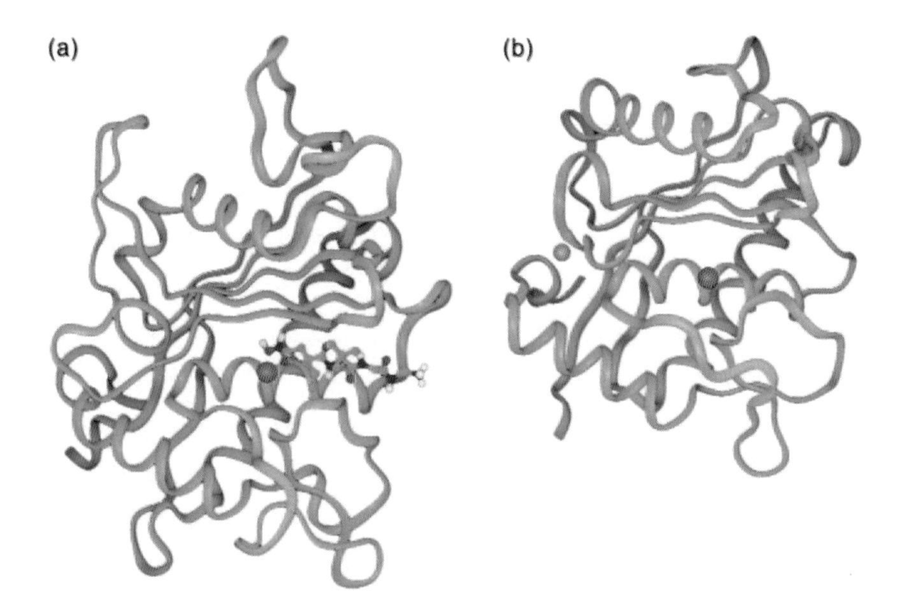


**Figure 3. X-ray structures of a TACE–inhibitor complex (a) and adamalysin II (b)**
(a) The protein backbone is shown in a 'worm' representation. Carbon (green), nitrogen (blue), oxygen (red) and hydrogen (white) atoms of the ligand, and the active-site zinc atom (purple) are shown [15]. (b) The zinc and calcium atoms are depicted as purple and white spheres, respectively [16].

highly conserved fold, with α-helices packed above and below a central β-sheet. Although the peripheral loops have different conformations, the core structure is generally similar to that of collagenase 1 and other MMP family members. The active-site cleft is delimited by β-strand 4, α-helix C and an additional protein chain from the C-terminal end of the catalytic domain. The substrate peptide chain is believed to bind to the cleft in an extended conformation, with an orientation that is antiparallel to that of β-strand 4. In this orientation, the substrate can make hydrogen bonds with β-strand 4 on the 'north' side of the cleft, and make additional hydrogen bonds with the protein on the 'south' side of the cleft, thereby effectively acting as an additional strand within an expanded β-sheet. TACE, adamalysin II, atrolysin C, and the MMPs all have a catalytic zinc atom positioned near the centre of the active-site cleft, where it can facilitate the cleavage of the scissile peptide bond. In all three ADAM-like proteases, the cleft has a deep lipophilic S1′ pocket to accommodate the lipophilic P1′ side-chain of the substrate (see Chapter 1, Figure 2). ADAM 10 is approx. 35% identical with TACE in the catalytic domain, which suggests that its catalytic domain should have a structure similar to that of TACE. The other ADAM family members have 25–40% identity with the SVMPs, suggesting that their overall structures should be similar to that of adamalysin II and atrolysin C.

## ADAM 10

ADAM 10, sometimes known as MADM, was the first ADAM family member that was shown to have a proteolytic function. ADAM 10 was purified from bovine brain based on its ability to process myelin basic protein [17]. Kuzbanian (*kuz*), the *Drosophila melanogaster* homologue of ADAM 10, was identified by screening for genes involved in neuronal development. The effects on neuronal development were subsequently linked to the cleavage of Notch, a Type I transmembrane receptor that controls cell fate determination in a broad spectrum of tissues (reviewed in [18]). Notch is thought to be activated by ligands displayed on, or possibly derived from, adjacent cells. The known ligands are Delta, Serrate and Lag 2, which are themselves Type I transmembrane proteins. Like APP, Notch is processed by different proteases at three different positions. The first cleavage reaction is performed by a furin-like protease, while the second processing event depends on *kuz* (in *Drosophila*) or ADAM 10 and/or TACE (in mammals), with the third and final cut being carried out within the membrane in a presenilin-dependent process. The first two cleavages are required to generate functional Notch, while the third cleavage occurs upon activation of Notch by its ligands, causing release of the cytosolic domain into the cytoplasm and subsequent translocation to the nucleus. The phenotype of the ADAM 10 knockout mice resembles the phenotype of the Notch knockout, supporting the idea that the second cleavage depends on ADAM 10 rather than TACE (reviewed in [2,11]).

As a further complication, ADAM 10 may also, or alternatively, cleave Delta, the Notch ligand.

As discussed above, there is evidence that ADAM 10 and TACE both function as α-secretases of APP. Transfection of HEK 293 cells with ADAM 10 increases the ability of the cells to produce sAPPα, providing further evidence that it is an α-secretase for APP processing. In addition, a construct of ADAM 10 carrying a point mutation in the catalytic domain, which renders it inactive, acts as a dominant negative factor, precluding release of sAPPα from transfected HEK 293 cells [19]. ADAM 10 can cleave a peptide containing the APP cleavage site, although the rate of cleavage is relatively low [20], as shown in Table 2. These results, together with the high levels of ADAM 10 found in the brain, suggest that ADAM 10 is an APP α-secretase.

Ephrin-A2, a well characterized axon repellant, appears to be a physiological substrate for ADAM 10 also [21]. Ephrin-A2 exists as a membrane bound form that binds Eph receptors on target cells. Ephrin-A2 binding can stimulate signalling through the tyrosine kinase domain of the receptor. However, after binding to the receptor, ephrin-A2 is processed by ADAM 10 to generate a soluble form. Since the soluble truncated form cannot activate the receptor, the cleavage process results in axon detachment and termination of signalling. A dominant negative mutant of ADAM 10 that lacked the protease domain was able to bind to ephrin-A2 and inhibit shedding of its ectodomain, and a cleavage-resistant mutant of ephrin-A2 led to greatly delayed axon withdrawal [21]. Interestingly, the *Drosophila* homologue of ADAM 10, *kuz*, was originally identified by genetic screening for proteins involved in normal axon extension.

Heparin-binding epidermal growth factor (HB-EGF), a member of the epidermal growth factor (EGF) family, which is important in cell transformation and mitogenesis, may also be a substrate for ADAM 10 [22]. Cleavage of the extracellular domain of HB-EGF generates a soluble fragment that activates the EGF receptor. In epithelial cells, this process can occur in response to lipoteichoic acid from Gram-positive bacteria, which stimulates the platelet-activating factor receptor. This receptor in turn activates ADAM 10, which leads to cleavage of HB-EGF [22]. This cleavage is part of the signalling pathway by which epithelial cells secrete mucin in response to Gram-positive bacteria. The same general ADAM-dependent pathway may be used by a variety of other G-protein-coupled receptors (GPCRs) to activate the EGF receptor in other cell types.

Considering the similarities between ADAM 10 and TACE, one might expect that ADAM 10 cleaves precursor TNF-α. However, as shown in Tables 1 and 2, the corresponding TNF-α peptide is cleaved much less efficiently by ADAM 10 than by TACE. ADAM 10 also processes precursor TNF-α inefficiently compared to TACE, and selective ADAM 10 inhibitors do not inhibit TNF release. Taken together, these data indicate that ADAM 10 is not a physiologically relevant TNF-α-converting enzyme.

**Table 2. Effect of recombinant ADAM 10 on peptide substrates**

An asterisk between the amino acids indicates where cleavage of the peptide substrate occurs. The dagger in the ephrin-A2 peptide indicates the predicted cleavage site. The specificity constant, $k_{cat}/K_m$, indicates how efficiently a particular substrate is cleaved by the enzyme; larger $k_{cat}/K_m$ values correspond to more efficient cleavage. Ac, acetyl; ND, not determined.

| Protein substrate | Peptide substrate | Specificity constant ($k_{cat}/K_m$) | Reference |
|---|---|---|---|
| Precursor TNF-α | SPLAQA*VRSSSR | $2 \times 10^3$ $M^{-1} \cdot s^{-1}$ | [20] |
| APP | YEVHHQK*LVFF | $7 \times 10^2$ $M^{-1} \cdot s^{-1}$ | [20] |
| Notch | Ac-CPYKIEA*VKSEPV | ND | |
| Myelin basic protein | YGSLP*QKAQRPQDEN | $9 \times 10^2$ $M^{-1} \cdot s^{-1}$ | [20] |
| Ephrin-A2 | YEAPEP†IFTSNS | ND | |

### ADAM 9

HB-EGF may also be a substrate for ADAM 9, a protease that is also referred to as meltrin-γ or MDC 9 (metalloproteinase disintegrin and cysteine-rich). Izumi et al. [9] showed that overexpression of full length ADAM 9 increases HB-EGF shedding. Mutation of the HEXXHXXGXXH zinc-binding motif in ADAM 9 led to apparent dominant negative effects, although this mutation also leads to a defect in folding and transport of the protein to the cell surface, thereby complicating the interpretation. Therefore, it is unclear if these results implicate ADAM 9 as a HB-EGF-processing enzyme. ADAM 9 can also process a peptide that spans the cleavage sequence of APP, and a classical MMP substrate [10]. ADAM 9 cleaves at a site distinct from the normal α-secretase site that generates sAPPα [10]. However, this alternative cleavage does actually occur in hippocampal neurons, suggesting that ADAM 9 could possibly be a hippocampal α-secretase. Interestingly, ADAM 9 is localized to the cell surface, where its MMP-like substrate specificity could enable it to degrade matrix components.

### ADAM 12

ADAM 12, or meltrin-α, an enzyme that is important for myoblast fusion, was identified at the same time as ADAM 9 by Yagami-Hiromasa et al. [23]. Two forms of ADAM 12 arise from alternative splicing: a soluble form, ADAM 12-S, and a long, membrane-bound form, ADAM 12-L. An α-2 macroglobulin assay was used to demonstrate that ADAM 12-S is proteolytically active [24]. A substrate for ADAM 12-S, insulin-like growth factor (IGF)-binding protein 3 (IGFBP-3), was then identified through a yeast two-hybrid system when placental proteins were screened for their ability to interact with IGFBP-3 [25]. IGFBP-3 is thought to function by sequestering IGF-I or -II until they are released by proteolysis of the binding protein through the action of ADAM 12-S and/or by other processing events. The balance between the IGFs and the IGFBPs is critical in osteoarthritis and diabetes, where decreased IGF levels are associated with the disease states. As discussed above for ADAM 10, ADAM 12 has also been implicated in the processing of HB-EGF in the GPCR/EGF receptor signalling pathway [26]. In this case, certain vasoactive molecules, such as phenylephrine, angiotensin II and endothelin-1, activate specific GPCRs in cardiomyocytes. These GPCRs induce ADAM 12 activity, which leads to shedding of HB-EGF and activation of the EGF receptor. Asakura et al. [26] found that a metalloprotease inhibitor could attenuate the cardiac hypertrophy normally induced by phenylephrine and angiotensin II.

## ADAM family proteases as targets in drug discovery

The membrane proteins that are subject to shedding by ADAM family proteases are involved in a number of serious diseases, and thus provide targets for drug discovery. TNF-α is again a convenient example, since it is responsible, in part, for the inflammation in RA. One approach to therapy involves TNF antagonists [27,28]. For example, Enbrel® (etanercept) is a TNF receptor fusion protein that binds TNF-α and effectively diverts some fraction of the endogenous TNF-α away from the TNF receptor on target cells. An Enbrel®/methotrexate combination has proven to be an effective treatment for RA [27]. Remicade® (infliximab) is a monoclonal antibody targeted against TNF-α, which is effective in the treatment of Crohn's disease, and in combination with methotrexate, is effective in RA [28]. These protein-based therapeutics are difficult to manufacture and deliver, and small molecule antagonists that could be delivered by mouth would be preferable. While it is generally difficult to obtain small molecules that directly antagonize cytokines, it is feasible to find small molecules that inhibit TACE, thereby preventing the secretion of the soluble, mature form of TNF-α. Non-selective TACE inhibitors have already been obtained from earlier MMP drug discovery projects, and have proven to be effective in preclinical models of septic shock and arthritis (reviewed in [2]). Several drug companies are working towards more selective compounds. Dupont is apparently the most successful to date, with a TACE inhibitor entering clinical trials.

## Conclusion

Many membrane proteins have extracellular domains that are subject to proteolytic release into the extracellular medium, a process known as shedding. Among the proteins subject to shedding are precursor TNF-α, TGF-α, the TNF receptors, APP and HB-EGF, which are thought to be involved in various disease processes such as arthritis, Crohn's disease, Alzheimer's disease and cardiac hypertrophy. Recent work has shown that ADAM family proteases are responsible for several of these shedding events, and evidence for their involvement in other shedding events is accumulating. ADAM family proteases are good targets for drug design, and inhibition of specific ADAM-mediated shedding processes may ultimately prove to be an effective therapy for a range of diseases.

### Summary

- *Many membrane proteins have extracellular domains that are subject to proteolytic release, a process known as 'shedding'.*
- *ADAMs are membrane-tethered proteases responsible for many shedding events.*
- *ADAM family proteases generally have a pro-domain, a catalytic domain, a disintegrin domain, a cysteine-rich domain, a transmembrane segment and a cytoplasmic tail.*

- *TACE (ADAM 17) is implicated in the shedding of TNF-α, TNF receptor I and TNF receptor II, TGF-α, APP, Notch, L-selectin and the interleukin-6 receptor.*
- *ADAM 10 is implicated in the shedding of Notch, APP, ephrin-A2 and HB-EGF.*
- *ADAM 9 has a unique substrate specificity, and may be involved in the processing of APP and HB-EGF.*
- *ADAM 12 has been implicated in the processing of HB-EGF and IGFBP.*
- *ADAM family proteases may be good targets for the development of novel therapies for a range of diseases.*

## References


1. Hooper, N.M., Karran, E.H. & Turner, A.J. (1997) Membrane protein secretases. *Biochem. J.* **321,** 265–279
2. Moss, M.L., White, J.M., Lambert, M.H. & Andrews, R.C. (2001) TACE and other ADAM proteases as targets for drug discovery. *Drug Discovery Today* **6,** 417–426
3. Black, R.A., Rauch, C.T., Lozlosky, C.J., Peschon, J.J., Slack, J.L., Wolfson, M.F., Castner, B.J., Stocking, K.L., Reddy, P., Srinivasan, S. et al. (1997). A metalloproteinase disintegrin that releases tumour-necrosis factor-α from cells. *Nature (London)* **385,** 729–732
4. Moss, M.L., Jin, S.-L.C., Milla, M.E., Burkhart, W., Carter, H.L., Chen, W.-J., Clay, W.C., Didsbury, J.R., Hassler, D., Hoffman, C.R. et al. (1997). Cloning of a disintegrin metalloproteinase that processes precursor tumor-necrosis factor-α. *Nature (London)* **385,** 733–736
5. Wolfsberg, T.G., Straight, P.D., Gerena, R.L., Huovila, A.P., Primakoff, P., Myles, D.G. & White, J.M. (1995) ADAM, a widely distributed and developmentally regulated gene family encoding membrane proteins with a disintegrin and metalloprotease domain. *Dev. Biol.* **169,** 378–383
6. Bode, W., Grams, F., Reinemer, P., Gomis-Ruth, F.X., Baumann, U., McKay, D.B. & Stocker, W. (1996) The metzincin-superfamily of zinc-peptidases. *Adv. Exp. Med. Biol.* **389,** 1–11
7. White, J.M., Bigler, D., Chen, M., Takahashi, Y. & Wolfsberg, T.G. (2001) ADAMs in cell adhesion. In *Cell Adhesion: Frontiers in Molecular Biology* (Beckerle, M., ed.), pp. 189–216, Oxford University Press, Oxford,
8. Jia, L.G., Shimokawa, K., Bjarnason, J.B. & Fox, J.W. (1996) Snake venom metalloproteinases: structure, function and relationship to the ADAMs family of proteins. *Toxicon* **34,**1269–1276
9. Izumi, Y., Hirata, M., Hasuwa, H., Iwamoto, R., Umata, T., Miyado, K., Tamai, Y., Kurisaki, T., Sehara-Fujisawa, A., Ohno, S. and Mekada, E. (1998) A metalloprotease-disintegrin MDC9/meltrin-gamma/ADAM9and PKCdelta are involved in TPA-induced ectodomain shedding of membrane-anchored heparin-binding EGF-like growth factor. *EMBO J.* **17,** 7260–7272
10. Roghani, M., Becherer, J.D., Moss, M.L., Atherton, R.E., Erdjument-Bromage, H., Arribas, J., Blackburn, R.K., Weskamp, G., Tempst, P. & Blobel, C.P. (1999) Metalloprotease-disintegrin MDC9: intracellular maturation and catalytic activity. *J. Biol. Chem.* **274,** 3531–3540
11. Schlondorff, J. & Blobel, C.P. (1999) Metalloprotease-disintegrins: modular proteins capable of promoting cell-cell interactions and triggering signals by protein-ectodomain shedding. *J. Cell Sci.* **112,** 3603-3617
12. Peschon, J.J., Slack, J.L., Reddy, P., Stocking, K.L., Sunnarborg, S.W., Lee, D.C., Russell, W.E., Castner, B.J., Johnson, R.S., Fitzner, J.N. et al. (1998) An essential role for ectodomain shedding in mammalian development. *Science* **282,** 1279–1280

13. Buxbaum, J.D., Liu, K.N., Luo, Y., Slack, J.L., Stocking K.L., Peschon, J.J., Johnson, R.S., Castner, B.J., Cerretti, D.P. & Black, R.A. (1998) Evidence that tumor necrosis factor alpha converting enzyme is involved in regulated alpha-secretase cleavage of the Alzheimer amyloid protein precursor. *J. Biol. Chem.* **273,** 27765–27767
14. Nunan, J. & Small, D.H. (2000) Regulation of APP cleavage by alpha-, beta- and gamma-secretases. *FEBS Lett.* **483,** 6–10
15. Maskos, K., Fernandez-Catalan, C., Huber, R., Bourenkov, G.P., Bartunik, H., Ellestad, G.A., Reddy, P., Wolfson, M.F., Rauch, C.G., Castner, B.J. et al. (1998) Crystal structure of the catalytic domain of human tumor necrosis factor-a-converting enzyme. *Proc. Natl. Acad. Sci. U.S.A.* **95,** 3408–3412
16. Gomis-Ruth, F.X., Kress, L.F., Kellermann, J., Mayr, I., Lee, X., Huber, R. & Bode, W. (1994) Refined 2.0 Å X-ray crystal structure of the snake venom zinc endopeptidase adamalysin II. Primary and tertiary structure determination, refinement, molecular structure and comparison with astacin, collagenase and thermolysin. *J. Mol. Biol.* **239,** 513–544
17. Howard, L. & Glynn, P. (1995). Membrane-associated metalloproteinase recognized by characteristic cleavage of myelin basic protein: assay and isolation. *Methods Enzymol.* **248,** 388–395
18. Kopan, R. & Goate, A. (2000) A common enzyme connects notch signaling and Alzheimer's disease. *Genes Dev.* **14,** 2799–2806
19. Lammich, S., Kojro, E., Postina, R., Gilbert, S., Pfeiffer, R., Jasionowski, M., Haass, C. & Fahrenholz, F. (1999) Constitutive and regulated alpha-secretase cleavage of Alzheimer's amyloid precursor protein by a disintegrin metalloprotease. *Proc. Natl. Acad. Sci. U.S.A.* **96,** 3922–3927
20. Amour, A., Knight, C.G., Webster, A., Slocombe, P.M., Stephens, P.E., Knauper, V., Docherty, A.J. & Murphy, G. (2000) The in vitro activity of ADAM-10 is inhibited by TIMP-1 and TIMP-3. *FEBS Lett.* **473,** 275–279
21. Hattori, M., Osterfield, M. & Flanagan, J.G. (2000) Regulated cleavage of a contact-mediated axon repellent. *Science* **289,** 1360–1365
22. Lemjabbar, H. & Basbaum, C. (2002) Platelet-activating factor receptor and ADAM10 mediate responses to Staphylococcus aureus in epithelial cells. *Nat. Med.* **8,** 41–46
23. Yagami-Hiromasa, T., Sato, T., Kurisaki, T., Kamijo, K., Nabeshima, Y. & Fujisawa-Sehara, A. (1995) A metalloprotease-disintegrin participating in myoblast fusion. *Nature (London)* **377,** 652–656
24. Loechel, F., Gilpin, B.J., Engvall, E., Albrechtsen, R. & Wewer, U.M. (1998) Human ADAM 12 (meltrin alpha) is an active metalloprotease. *J. Biol. Chem.* **273,** 16993–16997
25. Shi, Z., Xu, W., Loechel, F., Wewer, U.M. & Murphy, L.J. (2000) ADAM 12, a disintegrin metalloprotease, interacts with insulin-like growth factor-binding protein-3. *J. Biol. Chem.* **275,** 18574–18580
26. Asakura, M., Kitakaze, M., Takashima, S., Liao, Y., Ishikura, F., Yoshinaka, T., Ohmoto, H., Node, K., Yoshino, K., Ishiguro, H. et al. (2002) Cardiac hypertrophy is inhibited by antagonism of ADAM12 processing of HB-EGF: metalloproteinase inhibitors as a new therapy. *Nat. Med.* **8,** 35–40
27. Weinblatt, M.E., Kremer, J.M., Bankhurst, A.D., Bulpitt, K.J., Fleischmann, R.M., Fox, R.I., Jackson, C.G., Lange, M. & Burge, D.J. (1999) A trial of etanercept, a recombinant tumor necrosis factor receptor: Fc fusion protein, in patients with rheumatoid arthritis receiving methotrexate. *N. Engl. J. Med.* **340,** 253–259
28. Lipsky, P.E., van der Heijde, D.M., St Clair, E.W., Furst, D.E., Breedveld, F.C., Kalden, J.R., Smolen, J.S., Weisman, M., Emery, P., Feldmann, M. et al. (2000) Infliximab and methotrexate in the treatment of rheumatoid arthritis. *N. Engl. J. Med.* **343,** 1594–1602

# 12

# Regulated intramembrane proteolysis: from the endoplasmic reticulum to the nucleus

## Robert B. Rawson[1]

*Department of Molecular Genetics, University of Texas Southwestern Medical Center, 5323 Harry Hines Boulevard, Dallas, TX 75390-9046, U.S.A.*

### Abstract

Regulated intramembrane proteolysis (Rip) is an ancient and widespread process by which cells transmit information from one compartment (the endoplasmic reticulum) to another (the nucleus). Two separate cleavages that are carried out by two separate proteases are required for Rip. The first protease cleaves its protein substrate within an extracytoplasmic domain; the second cleaves it within a membrane-spanning domain, releasing a functionally active fragment of the target protein. In eukaryotes, examples of Rip can be divided into two classes, according to the proteases that are involved and the orientation of the substrates with the membrane. Class 1 Rip involves type 1 transmembrane proteins and requires presenilin for cleavage within a membrane-spanning domain. In Class 2 Rip, the highly hydrophobic metalloprotease, site-2 protease, is required for cleavage within a membrane-spanning domain and substrates are type 2 transmembrane proteins. Both classes of Rip are implicated in diseases that are important in modern societies, such as hyperlipidaemias (via the sterol

[1]*E-mail: rawson@utsw.swmed.edu*

regulatory element binding protein pathway) and Alzheimer's disease (via processing of the amyloid precursor protein.)

## Introduction

In an ever-changing world, survival depends on the ability to assess quickly and respond appropriately to changes in the environment. At the cellular level, an appropriate response often includes altering the complement of cellular proteins, as new protein functions are required while old ones become unnecessary. Synthesis of macromolecules is a time-consuming process that requires minutes or hours to go from inactive gene to transcript to functional protein. One strategy employed to ensure a cell's rapid response to environmental cues is the availability of previously synthesized transcriptional activators that are present outside the nucleus in an inactive form. Upon receipt of a specific signal, these pre-existing proteins are then activated and translocate to the nucleus where they direct the increased transcription of target genes. Many variations on this strategy have been identified and the activating events can include phosphorylation as in the case of the signal transducers and activators of transcription (STAT) factors. Signalling via STATs employs a mechanism whereby the latent, cytoplasmic transcription factors are phosphorylated by Janus family tyrosine kinases. Phosphorylation promotes dimerization of STATs and nuclear translocation, whereupon they modulate the expression of target genes [1]. In a different approach to keeping a pool of pre-existing factors out of the nucleus until needed, nuclear factor κB (NFκB)/Rel transcription factors are kept latent in the cytoplasm by interaction with the protein inhibitory κBα (IκBα). Receipt of an appropriate external signal causes IκBα to become phosphorylated, ubiquitinated, and then degraded by the proteasome (see Chapter 5). The free NFκB then translocates to the nucleus to activate its gene targets [2].

Recent investigations into several different systems have identified an unexpected role for proteolysis in the direct activation of pools of latent transcriptional activators that reside within the membrane compartment of the cell. These precursor proteins undergo proteolytic processing by two distinct enzymes in response to an intracellular signal. Different pairs of proteases are utilized to process different transcription factor precursors. The activity of the first enzyme in each pair is directly regulated, while the activity of the second enzyme is regulated indirectly by substrate availability dependent on the first, regulated cleavage. The second enzymes are particularly interesting because they cleave substrates within domains that are believed normally to lie within the plane of the membrane. In light of these observations, this phenomenon has been termed regulated intramembrane proteolysis (Rip) [3].

## New appreciation for an old strategy

Rip arose early in evolution, with examples being found in organisms from prokaryotes to primates. Processes regulated by Rip range from control of metabolism to differentiation and development. In the bacterium *Bacillus subtilis*, Rip controls the differentiation of the endospore, a dormant form of the organism formed in response to nutrient deprivation. During this process, Rip-mediated release of the membrane-bound, transcriptional regulatory protein pro-$\sigma^{K}$ is necessary to complete the programme of sporulation [4]. In another prokaryote, *Enterococcus*, Rip is required to produce a pheromone that is needed to initiate aggregation behaviour [3]. The bacterial proteases involved in the generation of active transcription factors and signalling molecules are encoded by the genes SpoIVFB in *Bacillus* and eep in *Enterococcus* and are related in sequence to mammalian enzymes that catalyse the second cleavage event of Rip.

In eukaryotes, examples of Rip fall into two distinct classes (Table 1). Class 1 Rip involves type-1 transmembrane proteins (N-terminus extracytoplasmic, C-terminus cytoplasmic) whose cleavage is dependent upon γ-secretase activity. This activity requires two or more proteins acting either as the γ-secretase or as cofactors. Genetic and biochemical evidence indicates that γ-secretase activity involves the presenilin polytopic membrane proteins [5] and the transmembrane glycoprotein nicastrin [6]. Proteolysis appears to take place at or near the plasma membrane. Proteins known to undergo Class 1 Rip include the developmental signalling molecule Notch [7], the human epidermal growth factor receptor-like protein ErbB-4 [8], and the amyloid precursor protein (APP) (the proteolysis of APP is discussed in more detail in Chapter 4).

Intramembranous cleavage of APP came to attention owing to the fact that one of the products of this cleavage, the amyloid-β peptide, forms the amyloid plaques that are believed to cause neuronal death in Alzheimer's disease [9]. Recent investigation into the function of the cytoplasmic domain of APP supports the notion that Rip is a mechanism for controlling gene expression. This domain of APP forms a multimeric complex composed of the nuclear adaptor protein Fe65 and the histone acetyltransferase Tip60 [10]. This complex in turn stimulates mRNA synthesis via interaction with transcription factors that have DNA-binding domains. While the latter proteins have not yet been identified, this result indicates that proteolytic release of the cytoplasmic domain of APP by Rip may regulate gene expression [10].

Recent results indicate that some instances of intramembranous cleavage of type-1 transmembrane proteins do not require γ-secretase and do not release a transcription factor domain to the cytosol [11,12]. In *Drosophila*, soluble Spitz (a protein homologous with vertebrate transforming growth factor α) is a ligand for the *Drosophila* epidermal growth factor receptor. Spitz is synthesized as a type-1 transmembrane protein with the ligand portion facing the lumen of the endoplasmic reticulum (ER). Genetic evidence indicates that processing of Spitz depends

**Table 1. Eukaryotic proteins that undergo Rip**

*For APP and ErbB-4, the *in vivo* targets of transcriptional activation are unknown, although a role for ErbB-4 in growth and differentiation has been suggested. †It is not known whether nicastrin is required for cleavage of ErbB-4 or which of the presenilins may be normally involved. For further discussion of these proteins and for a brief description of a potentially new class of intramembrane proteolysis involving the Star–Spitz–Rhomboid-1 pathway in *Drosophila*, see text.

| Protein product | Membrane orientation | Proteins required for intramembrane cleavage | Function of released fragment | References |
|---|---|---|---|---|
| Class 1 | | | | |
| Notch | Type 1 | Presenilins, nicastrin | Transcriptional regulation (growth and differentiation) | [5–7] |
| APP | Type 1 | Presenilins, nicastrin | Transcriptional regulation* | [9,10] |
| ErbB-4 | Type 1 | Presenilins† | Transcriptional regulation* (growth and differentiation) | [8] |
| Class 2 | | | | |
| SREBP-1 and -2 | Type 2 | S2P | Transcriptional regulation (lipid metabolism) | [3,13] |
| ATF6$\alpha$ and ATF6$\beta$ | Type 2 | S2P | Transcriptional regulation (ER stress response) | [3,26] |

on both Star and Rhomboid-1 proteins. Lee et al. [11] demonstrated that the role of Star in Spitz processing is to escort newly-synthesized Spitz from the ER to the Golgi. At the Golgi, Rhomboid-1 is required for the proteolytic release of soluble Spitz into the lumen of the ER and its subsequent secretory pathway. Urban, Lee and Freeman [12] provide evidence that Spitz is cleaved within its transmembrane domain and that Rhomboid-1 may be the prototype for a novel family of serine proteases involved in this intramembrane proteolysis. The Star–Spitz–Rhomboid-1 pathway therefore points to expanded roles for intramembrane proteolysis in eukaryotes, beyond the direct control of transcription.

Proteins involved in Class 2 Rip in eukaryotes are type-2 transmembrane proteins (N-terminus cytoplasmic, C-terminus extracytoplasmic) (Figure 1a). The best understood example of Class 2 Rip is the membrane-bound transcription factor sterol regulatory element binding protein (SREBP). The first cleavage of SREBP produces a type-2 transmembrane protein that is the substrate for the second protease. Recently, proteolytic release of another type-2 mem-

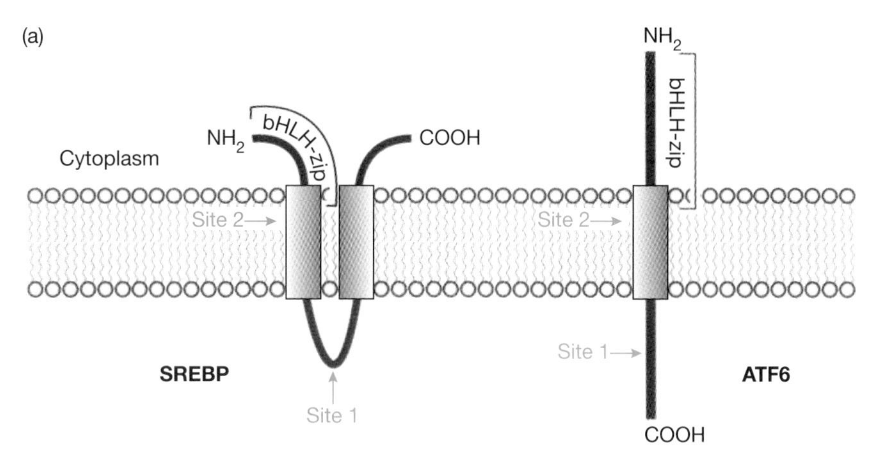


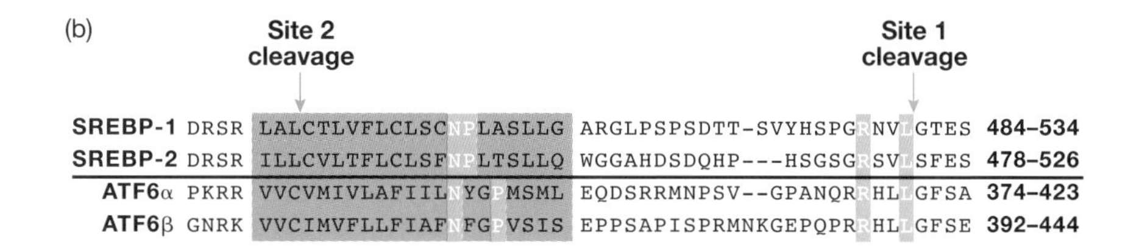


**Figure 1. Proteins processed by Class 2 Rip**

(a) Schematic diagram of the orientation of SREBP and ATF6 within the membrane of the ER. The sites of cleavage and the position of the basic helix-loop-helix leucine zipper (bHLH-zip) transcription factor domains are indicated. (b) Sequence comparison of the membrane-spanning and luminal domains of SREBP and ATF6. The positions of site 1 and site 2 cleavage are indicated. The membrane-spanning domain is indicated by grey shading. Residues that are critical for site 1 or site 2 cleavage of ATF6α [26] and SREBP-2 [27] are indicated by blue shading. Identical residues in SREBP-1 and ATF6β are also indicated. The GenBank accession numbers for the human protein sequences shown are: SREBP-1, P36956; SREBP-2, A54962; ATF6α, AAB64434; ATF6β, NP_004372.

brane-spanning transcription factor, activating transcription factor 6 (ATF6) (see below), has been shown to require the same proteases that process SREBP. These examples of eukaryotic Class 2 Rip are involved in transmitting signals from the ER to the nucleus. This essay focuses on the regulation of Class 2 Rip and the mechanisms by which control is accomplished.

## Class 2 Rip: the processing machinery

At present, three different loci are known to comprise the processing machinery required for Class 2 Rip in mammalian cells: SREBP cleavage activating protein (SCAP; Figure 2a), site-1 protease (S1P; Figure 2b) and site-2 protease (S2P; Figure 2c). The genes that encode these proteins were isolated by expression and complementation cloning using mammalian cells with defects in the feedback regulation of cholesterol biosynthesis owing to disrupted processing of SREBPs [13]. S1P and S2P were the first Rip proteases whose encoding cDNAs were cloned.

S1P (also known as SKI-1; subtilisin/kexin-like isoenzyme [14]) is a subtilisin-like, membrane-bound serine protease that is located in the ER and Golgi. It is a type 1 membrane protein, with the bulk of the protein, including the active site, in the lumen and only a short cytoplasmic C-terminal tail (Figure 2b). S1P contains a classic serine protease catalytic triad ($His^{218}$, $Asp^{249}$ and $Ser^{414}$) and, like other subtilases, it is synthesized as an inactive proenzyme that undergoes autocatalytic processing prior to activation [15]. Active S1P resides in the Golgi apparatus. S1P/SKI-1 is crucial for Rip involving ATF6 and SREBP as substrates. The protein may also play other roles in cellular physiology. It has been shown to cleave a number of other substrates and may be active when secreted into the medium (see Chapter 7).

S2P contains an HEXXH zinc-binding motif that is characteristic of several families of metalloproteases, although its sequence is otherwise unlike any previously described enzyme. The residues of the putative zinc-binding site are essential for S2P function; substitutions in the HEXXH signature sequence abolish its ability to restore SREBP cleavage in mutant cells that lack S2P [16]. The glutamate residue within this motif is thought to act as a nucleophile in peptide bond cleavage. In contrast with many other metalloproteases, the substitution of the glutamate residue with aspartate reduces, but does not eliminate, S2P activity. In a distant S2P homologue, the SpoIVFB protease of *Bacillus*, the analogous substitution also permits significant activity. These results emphasize one of several differences between S2P-like enzymes and other zinc-binding metalloproteases and render the proteins sufficiently distinct to be included as a separate subfamily (M50, [17]).

S2P is a hydrophobic protein with several putative membranous domains that are separated by distinct hydrophilic domains and are accessible to the glycosylation machinery of the ER and Golgi [18]. The hydrophobic regions

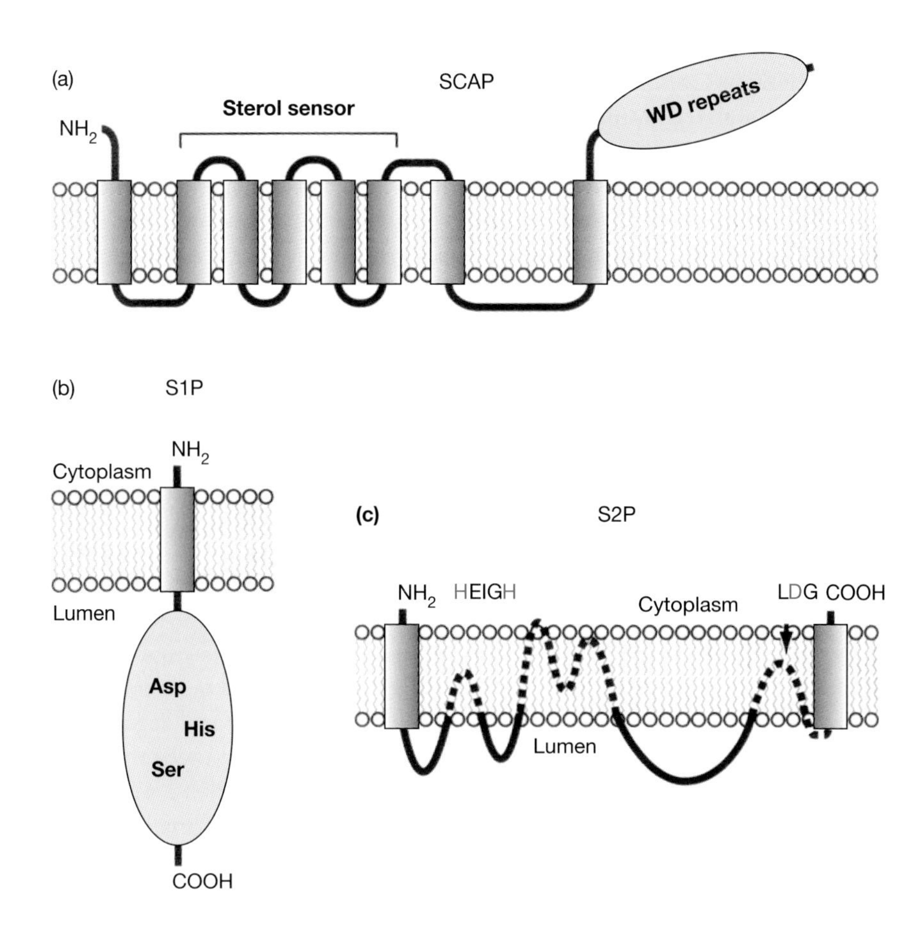


**Figure 2. Proteins required for Class 2 Rip in mammalian cells**
(a) SCAP contains eight membrane-spanning helices in its N-terminal domain. Helices 2–6 comprise the sterol sensor. The C-terminal domain contains WD repeats (so-called because of the presence of tryptophan and aspartate residues). The topology shown is based on the data of Nohturfft, Brown and Goldstein [32]. SCAP is required for cleavage of SREBPs but not for cleavage of ATF6. S1P and S2P are required for cleavage of both SREBPs and ATF6 [26]. (b) S1P contains a single membrane-spanning helix and a large luminal domain that includes residues comprising the catalytic triad of the active site of a serine protease (Asp, His, Ser) [13]. The luminal location of its active site is necessary as its substrates lie within the luminal domains of target proteins (see Figure 1a). (c) S2P is a highly hydrophobic, integral membrane protein. The indicated amino acid motifs (HEIGH and LDG) are essential for S2P function. Residues that serve as ligands for the active site zinc are indicated in blue. The topology is based on the data of Zelenski et al. [18].

that separate the hydrophilic domains do not span the membrane but appear to be embedded within it (Figure 2c).

The third component of the processing machinery of Class 2 Rip is the large polytopic membrane protein SCAP, which is necessary for cleavage of SREBPs [13]. Six of the eight N-terminal, membrane-spanning helices of SCAP comprise a sterol-sensing domain (Figure 2a). This designation rests on two lines of evidence. (1) Similar domains are found in a region of 3-hydroxy-3-methyl glutaryl-CoA reductase that controls sterol-regulated degradation of the protein. Related domains are also found in the gene that is defective in Niemann–Pick Type-C disease, a cholesterol storage disorder, and in the Patched and dispatched gene products of the hedgehog signalling pathway. The hedgehog protein is modified covalently with a cholesterol moiety [19]. (2) Four independently isolated mutations that reduce or abolish sterol regulation of SCAP function occur at two residues within the sterol-sensing domain [13].

The C-terminal portion of SCAP is located in the cytoplasm and consists of five copies of a motif designated WD repeats (owing to the presence of tryptophan and aspartate residues) that mediate protein–protein interactions [20]. In order for the first (site 1) cleavage event to occur within SREBP, SCAP and SREBP must form a complex via their C-termini. Formation of this complex is not regulated by sterols; however, the budding of the heterodimer into vesicles is [21]. The role of SCAP in Class 2 Rip appears to be limited to cleavage of SREBPs, as the molecule does not appear to be required for cleavage of ATF6.

## Rip and the control of lipid biosynthesis: SREBPs

There are three distinct SREBP proteins, the products of two separate genes, that play differential but overlapping roles in the transcriptional regulation of lipid metabolism. SREBP-1a and SREBP-1c mainly activate the transcription of genes that are required for fatty acid synthesis. SREBP-2 principally activates the transcription of genes that are required for cholesterol biosynthesis and uptake [22]. The SREBPs are synthesized as large precursors that are inserted into the membrane of the ER in a hairpin fashion, with both the N-terminal transcription factor domain and the C-terminal regulatory domain located within the cytoplasm. Two membrane-spanning domains, which are separated by a short luminal loop, anchor the precursor to the membrane and produce the hairpin orientation (Figure 1a).

SREBPs are cleaved in response to cellular demand for sterols, and the sterol-sensing domain of SCAP is central to this feedback regulation. When this domain of SCAP is overexpressed, sterols no longer regulate cleavage of SREBPs, the active form of which is then produced constitutively [23]. The sterol-sensing domain is thought to compete with full-length SCAP for interaction with an unidentified protein that normally retains it within the ER in the presence of sterols. When this interaction is blocked by a truncated sterol-sensing domain, the full-length SCAP–SREBP complex is free to travel from the ER to a post-ER/Golgi compartment in an unregulated fashion. This results in constitutive cleavage of SREBPs.

Once the SREBP–SCAP complex reaches the Golgi, active S1P cleaves SREBP within its luminal loop, releasing the N-terminal transcription factor domain from the C-terminal SCAP-binding domain. Both halves of the protein remain bound to the membrane owing to the presence of one membrane-spanning domain in each. This domain in the N-terminal half is the site of the second cleavage by S2P, which releases the active transcription factor from the membrane, thereby enabling movement to the nucleus and gene activation. The overlapping transcriptional specificity of SREBP-1 and SREBP-2, coupled with the processing of both proteins via Rip, allows animal cells to co-ordinate the biosynthesis of the principal membrane components, cholesterol and fatty acids.

## Rip and the unfolded protein response: ATF6

ATF6 (also known as ATF6α, [24]) is one of two closely related type-2 transmembrane proteins that are involved in the transcriptional response to ER stress (also known as the unfolded protein response). The other protein is ATF6β (formerly G13, [24]). Like the SREBPs, the cytoplasmic, N-terminal domains of ATF6α and ATF6β encode transcription factors of the basic helix-loop-helix leucine zipper family (Figure 1b). When unfolded proteins accumulate in the ER, transcription of genes encoding folding enzymes and chaperone proteins is upregulated owing to ATF6 binding to ER stress response elements in their promoters. ATF6 is also cleaved proteolytically to free the active factor from the membrane and allow movement to the nucleus [25]. Like SREBP, two separate cleavages are required for this release: the first, luminal cleavage largely depends on S1P, while the second cleavage requires S2P [26].

## Class 2 Rip: the substrates

The amino acid sequences recognized and cleaved by S1P at site 1 were first identified for SREBP-2 [13]. S1P cleaves the $Leu^{522}$–$Ser^{523}$ bond in the sequence RSVL↓S (Figure 1b). $Arg^{519}$ and $Leu^{522}$ are both important for recognition by S1P. When $Arg^{519}$ is replaced with alanine ($Arg^{519}$→Ala), cleavage at site 1 is abolished. Even the conservative substitution $Arg^{519}$→Lys greatly reduces the efficiency of cleavage. Similarly, the requirement for $Leu^{522}$ is very stringent; substitution of this residue by valine completely abolishes cleavage [27]. These same requirements are observed for cleavage at the $Leu^{419}$–$Gly^{420}$ bond in the sequence RHLL↓G in the luminal domain of ATF6α. The mutations $Arg^{416}$→Ala or $Leu^{419}$→Val abolish site 1 cleavage of ATF6α by S1P [26]. It is likely that ATF6β is also a substrate for S1P as it shares an identical luminal sequence ($R^{437}$HLLG) with ATF6α (Figure 1B).

In mammalian cells that lack S1P, processing of SREBPs cannot occur. As a result, these cells cannot upregulate transcription of the genes of cholesterol and fatty acid synthesis and are therefore auxotrophic for these lipids. Processing of ATF6α is also greatly reduced in these cells. Interestingly, when fed free cholesterol and unsaturated fatty acid, the mutant cells grow at a rate that is indistinguishable from that of their wild-type counterparts. Similarly, in cholesterol auxotrophic cells that lack S2P, no ER-stress-related growth or survival phenotype is observed under standard culture conditions, even though processing of ATF6α is completely abolished [26].

Studies of wild-type ATF6α in mutant cells that lack S2P confirm the involvement of this protease in the ER stress response. Both ATF6 cleavage and induced transcription of the ATF6 target gene, which encodes Bip (Grp78; glucose-regulated protein of 78 kDa), are profoundly deficient in cholesterol auxotrophic mutant cells lacking S2P [26]. These results are complemented by experiments with point mutants of ATF6α in wild-type cells. In order for a type 2 transmembrane protein to serve as a substrate for S2P, the substrate protein must have a luminal domain of no more than 25 amino acids and an asparagine/proline motif within its adjacent membrane-spanning helix [28]. Both SREBP and ATF6 are cleaved by S1P at a site C-terminal of the membrane-spanning domain (by 20–23 amino acids) (Figure 1b). The second requirement for asparagine or proline within the membrane-spanning domain is also observed in both SREBP and ATF6. When either the asparagine or proline residues in the first membrane-spanning domain of SREBP-2 are mutated to phenylalanine or leucine respectively, site 2 cleavage still occurs [28]. Simultaneous substitution of both residues, however, eliminates cleavage at site 2 and leads to the accumulation of the membrane-bound intermediate formed by site 1 cleavage [28]. This result holds for an asparagine/proline motif in the membrane-spanning domain of ATF6α as well ($N^{391}$YGP; Figure 1). Single substitutions of either $Asn^{391}$→Phe or $Pro^{394}$→Leu do not abolish cleavage of ATF6α by S2P. By contrast, the double substitution ($Asn^{391}$→Phe and $Pro^{394}$→Leu) completely blocks production of the nuclear form of ATF6α and leads to increased accumulation of a membrane-bound intermediate form [26]. Again, it seems likely that ATF6β shares these same requirements for processing, as the protein contains an $N^{410}$FGP motif in a membrane-spanning domain (Figure 1b).

The asparagine/proline motif is common at the ends of α-helices in globular proteins where it is thought to cap the N-termini [29], but it is rarely found in membrane-spanning domains. In a hydrophobic, membrane-spanning helix, the asparagine/proline motif may facilitate a transition from the usual α-helical conformation to a more extended conformation, which may be more susceptible to proteolytic attack by S2P than an α-helix would be [28]. Unfolding of the membrane-spanning α-helix in this scenario need not be an abrupt transition from a hydrophobic environment within the membrane to a hydrophilic one in the cytoplasm. Considering the unusually hydrophobic nature of S2P, some of its membranous sequences could conceivably stabilize the substrate in an extended conformation prior to proteolysis.

In contrast with cells that lack S1P or S2P, processing of ATF6α is normal in mutant cells that lack SCAP [26]. This conclusion is further supported by the fact that levels of cellular sterol that strongly induce or suppress cleavage of SREBPs have no effect on the processing of ATF6. Reciprocally, treatments that induce ER stress, and therefore ATF6 processing, have little effect on

SREBP cleavage. Since active S1P, and presumably S2P, resides in a post-ER/Golgi compartment, it is likely that ATF6 must also exit the ER in order to be processed. It is conceivable that an unidentified factor performs a function analogous to that of SCAP as an ER to Golgi escort for ATF6.

In both known examples of Class 2 Rip, a central feature of the process is the initial ER localization of the membrane-bound transcription factor. Release of the cytoplasmic domain depends on events in the ER that initiate transport of the substrate from the ER to another compartment where it can be cleaved. Thus, the S1P and S2P proteolytic machinery has been put into the service of communication between the ER (where sterols are synthesized and many proteins are folded) and the nucleus. By means of this seemingly roundabout mechanism, events in the ER result in alteration of a cell's transcriptional programme.

## Communication between the ER and nucleus without Rip: yeast

All prokaryotic and metazoan genomes that have been sequenced to date encode homologues of S2P. The chromosomes of higher plants also contain clear orthologues of both S1P and S2P, as revealed by database searches (R.B. Rawson, unpublished work). It thus seems reasonable to speculate that Rip may occur in this important group of eukaryotes as well, although no examples are yet reported. Given the widespread occurrence of S1P- and S2P-like genes, it is surprising that no examples of Rip (and no S2P homologues) are found in fungi.

An alternative strategy for the processing of membrane-spanning transcription factors occurs in fungi. This regulatory pathway is termed regulated ubiquitin/proteasome-dependent processing (RUP; [30].) In the yeast *Saccharomyces cerevisiae*, the membrane-bound transcription factors SPT23 and MGA2 (which are NFκB homologues) are processed in a ubiquitin-dependent manner to release a transcriptionally active fragment to the nucleus. Intriguingly, SPT23 and MGA2 are also necessary for the production of unsaturated fatty acids in yeast [30]. Having membrane-spanning transcription factors controlling aspects of lipid metabolism in response to specific proteolysis of precursors makes RUP logically similar to Rip; however, the two processes differ in their mechanisms. RUP depends on component proteases of the proteasome rather than a pair of membrane-bound proteases. Perhaps yeast, having no S2P-like enzymes, employ a different mechanism for transmitting information about the lipid status of the cell from the ER to the nucleus.

## Rip in disease

Examples of both Class 1 and Class 2 Rip play or may play central roles in diseases that afflict large numbers of people in Western societies. For example, misregulation of SREBP cleavage and the consequent overproduction of lipids contribute to hyperlipidaemias and, therefore, to vascular disease, strokes and heart attacks. Such misregulation may also contribute to fatty liver pathologies.

On the other hand, the unfolded protein response can be triggered by many different disease states, including viral infection, in which the synthesis of viral proteins overwhelms the ER folding machinery. Disruptions in the ER stress response may thus be involved in the pathology of such infections.

For Class 1 Rip, the most dramatic disease known is Alzheimer's. It is likely that this debilitating neurodegenerative condition results from the deposition of amyloid plaques, which are composed of a direct product of Rip [3]. Some of the mutations that are responsible for inherited forms of early onset Alzheimer's disease are point mutations within APP itself. The majority, however, are mutations in the presenilins that are required for intramembranous cleavage of substrates in Class 1 Rip. All of these mutations result in increased production of the amyloidogenic amyloid $\beta_{1-42}$ fragment [31] (see Chapter 4). This biochemical phenotype indicates the crucial role of Rip in a very common disease. More broadly, the role of the Notch signalling pathway during development suggests the plausible hypothesis that perturbations in Class 1 Rip could be a cause of certain birth defects.

## Conclusions

Appreciation of Rip is recent, many details of the phenomenon are incompletely understood and some intriguing questions remain unanswered. For example, how did such an involved system come into being? Why are two cleavages necessary when it would appear that one would suffice? What other proteins are involved? Because of a proven or potential involvement in common diseases, it is clear that understanding more about the mechanisms of Rip will contribute to our knowledge of both cellular signalling and pathological states.

### Summary

- *Rip is an ancient and widespread process for controlling gene expression.*
- *Two separate cleavages by two separate proteases are required, one in the extracytoplasmic space and one within a membrane-spanning domain.*
- *In eukaryotes, Rip falls into two classes that are defined by the enzymes involved and the nature of their substrates*
- *Class 1 Rip depends on a complex containing presenilin for intramembranous cleavage. Class 2 Rip depends on an integral membrane metalloprotease, S2P.*

- *Both classes of Rip play crucial roles in diseases common in modern societies such as hyperlipidaemias and atherosclerosis (via the SREBP pathway) and Alzheimer's disease (via APP processing.)*

I am grateful to David Russell for critically reading this manuscript. The author is supported by grants from the National Institutes of Health (HL20948) and the American Heart Association (0130010N).

## References


1. Leonard, W.J. & O'Shea, J.J. (1998) Jaks and STATs: biological implications. *Annu. Rev. Immunol.* **16**, 293–322
2. Ghosh, S., May, M.J. & Kopp, E.B. (1998) NF-kappa B and Rel proteins: evolutionarily conserved mediators of immune responses. *Annu. Rev. Immunol.* **16**, 225–260
3. Brown, M.S., Ye, J., Rawson, R.B. & Goldstein, J.L. (2000) Regulated intramembrane proteolysis: a control mechanism conserved from bacteria to humans. *Cell* **100**, 391–398
4. Rudner, D.Z., Fawcett, P. & Losick, R. (1999) A family of membrane-embedded metalloproteases involved in regulated proteolysis of membrane-associated transcription factors. *Proc. Natl. Acad. Sci. U.S.A.* **96**, 14765–14770
5. Li, Y.M., Lai, M.T., Xu, M., Huang, Q., DiMuzio-Mower, J., Sardana, M.K., Shi, X.P., Yin, K.C., Shafer, J.A. & Gardell, S.J. (2000) Presenilin 1 is linked with gamma-secretase activity in the detergent solubilized state. *Proc. Natl. Acad. Sci. U.S.A.* **97**, 6138–6143
6. Yu, G., Nishimura, M., Arawaka, S., Levitan, D., Zhang, L., Tandon, A., Song, Y.Q., Rogaeva, E., Chen, F., Kawarai, T. et al. (2000) Nicastrin modulates presenilin-mediated notch/glp-1 signal transduction and betaAPP processing. *Nature (London)* **407**, 48–54
7. De Strooper, B., Annaert, W., Cupers, P., Saftig, P., Craessaerts, K., Mumm, J.S., Schroeter, E.H., Schrijvers, V., Wolfe, M.S., Ray, W.J. et al. (1999) A presenilin-1-dependent gamma-secretase-like protease mediates release of Notch intracellular domain. *Nature (London)* **398**, 518–522
8. Ni, C.Y., Murphy, M.P., Golde, T.E. & Carpenter, G. (2001) gamma-Secretase cleavage and nuclear localization of ErbB-4 receptor tyrosine kinase. *Science* **294**, 2179–2181
9. Selkoe, D.J. (1997) Alzheimer's disease: genotypes, phenotypes, and treatments. *Science* **275**, 630–631
10. Cao, X. & Sudhof, T.C. (2001) A transcriptively active complex of APP with Fe65 and histone acetyltransferase Tip60. *Science* **293**, 115–120
11. Lee, J.R., Urban, S., Garvey, C.F. & Freeman, M. (2001) Regulated intracellular ligand transport and proteolysis control EGF signal activation in Drosophila. *Cell* **107**, 161–171
12. Urban, S., Lee, J.R. & Freeman, M. (2001) Drosophila rhomboid-1 defines a family of putative intramembrane serine proteases. *Cell* **107**, 173–182
13. Brown, M.S. & Goldstein, J.L. (1999) A proteolytic pathway that controls the cholesterol content of membranes, cells, and blood. *Proc. Natl. Acad. Sci. U.S.A.* **96**, 11041–11048
14. Seidah, N.G., Mowla, S.J., Hamelin, J., Mamarbachi, A.M., Benjannet, S., Toure, B.B., Basak, A., Munzer, J.S., Marcinkiewicz, J., Zhong, M. et al. (1999) Mammalian subtilisin/kexin isozyme SKI-1: a widely expressed proprotein convertase with a unique cleavage specificity and cellular localization. *Proc. Natl. Acad. Sci. U.S.A.* **96**, 1321–1326
15. Espenshade, P.J., Cheng, D., Goldstein, J.L. & Brown, M.S. (1999) Autocatalytic processing of site-1 protease removes propeptide and permits cleavage of sterol regulatory element-binding proteins. *J. Biol. Chem.* **274**, 22795–22804
16. Rawson, R.B., Zelenski, N.G., Nijhawan, D., Ye, J., Sakai, J., Hasan, M.T., Chang, T.Y., Brown, M.S. & Goldstein, J.L. (1997) Complementation cloning of S2P, a gene encoding a putative metalloprotease required for intramembrane cleavage of SREBPs. *Mol. Cell* **1**, 47–57

17. Rawlings, N.D. & Barrett, A.J. (1999) MEROPS: the peptidase database. *Nucleic Acids Res.* **27**, 325–331
18. Zelenski, N.G., Rawson, R.B., Brown, M.S. & Goldstein, J.L. (1999) Membrane topology of S2P, a protein required for intramembranous cleavage of sterol regulatory element-binding proteins. *J. Biol. Chem.* **274**, 21973–21980
19. Porter, J.A., Young, K.E. & Beachy, P.A. (1996) Cholesterol modification of hedgehog signaling proteins in animal development. *Science* **274**, 255–259
20. Neer, E.J., Schmidt, C.J., Nambudripad, R. & Smith, T.F. (1994) The ancient regulatory protein family of WD-repeat proteins. *Nature (London)* **371**, 297–300
21. Nohturfft, A., Yabe, D., Goldstein, J.L., Brown, M.S. & Espenshade, P.J. (2000) Regulated step in cholesterol feedback localized to budding of SCAP from ER membranes. *Cell* **102**, 315–323
22. Horton, J.D. & Shimomura, I. (1999) Sterol regulatory element-binding proteins: activators of cholesterol and fatty acid biosynthesis. *Curr. Opin. Lipidol.* **10**, 143–150
23. Yang, T., Goldstein, J.L. & Brown, M.S. (2000) Overexpression of membrane domain of SCAP prevents sterols from inhibiting SCAP·SREBP exit from endoplasmic reticulum. *J. Biol. Chem.* **275**, 29881–29886
24. Haze, K., Okada, T., Yoshida, H., Yanagi, H., Yura, T., Negishi, M. & Mori, K. (2001) Identification of the G13 (cAMP-response-element-binding protein-related protein) gene product related to activating transcription factor 6 as a transcriptional activator of the mammalian unfolded protein response. *Biochem. J.* **355**, 19–28
25. Haze, K., Yoshida, H., Yanagi, H., Yura, T. & Mori, K. (1999) Mammalian transcription factor ATF6 is synthesized as a transmembrane protein and activated by proteolysis in response to endoplasmic reticulum stress. *Mol. Biol. Cell* **10**, 3787–3799
26. Ye, J., Rawson, R.B., Komuro, R., Chen, X., Dave, U.P., Prywes, R., Brown, M.S. & Goldstein, J.L. (2000) ER stress induces cleavage of membrane-bound ATF6 by the same proteases that process SREBPs. *Mol. Cell* **6**, 1355–1364
27. Duncan, E.A., Brown, M.S., Goldstein, J.L. & Sakai, J. (1997) Cleavage site for sterol-regulated protease localized to a leu-Ser bond in the lumenal loop of sterol regulatory element-binding protein-2. *J. Biol. Chem.* **272**, 12778–12785
28. Ye, J., Dave, U.P., Grishin, N.V., Goldstein, J.L. & Brown, M.S. (2000) Asparagine-proline sequence within membrane-spanning segment of SREBP triggers intramembrane cleavage by site-2 protease. *Proc. Natl. Acad. Sci. U.S.A.* **97**, 5123–5128
29. Richardson, J.S. & Richardson, D.C. (1988) Amino acid preferences for specific locations at the ends of alpha helices. *Science* **240**, 1648–1652
30. Hoppe, T., Matuschewski, K., Rape, M., Schlenker, S., Ulrich, H.D. & Jentsch, S. (2000) Activation of a membrane-bound transcription factor by regulated ubiquitin/proteasome-dependent processing. *Cell* **102**, 577–586
31. Esler, W.P. & Wolfe, M.S. (2001) A portrait of Alzheimer secretases – new features and familiar faces. *Science* **293**, 1449–1454
32. Nohturfft, A., Brown, M.S. & Goldstein, J.L. (1998) Topology of SREBP cleavage-activating protein, a polytopic membrane protein with a sterol-sensing domain. *J. Biol. Chem.* **273**, 17243–17250

# 13

# Protease-activated receptors: the role of cell-surface proteolysis in signalling

Graeme S. Cottrell, Anne-Marie Coelho and Nigel W. Bunnett[1]

*Department of Surgery, University of California at San Francisco, 521 Parnassus Avenue, San Francisco, CA 94143-0660, U.S.A., and Department of Physiology, University of California at San Francisco, 521 Parnassus Avenue, San Francisco, CA 94143-0660, U.S.A.*

## Abstract

Certain extracellular proteases, derived from the circulation and inflammatory cells, can specifically cleave and trigger protease-activated receptors (PARs), a small, but important, sub-group of the G-protein-coupled receptor super-family. Four PARs have been cloned and they all share the same basic mechanism of activation: proteases cleave at a specific site within the extracellular N-terminus to expose a new N-terminal tethered ligand domain, which binds to and thereby activates the cleaved receptor. Thrombin activates PAR1, PAR3 and PAR4, trypsin activates PAR2 and PAR4, and mast cell tryptase activates PAR2 in this manner. Activated PARs couple to signalling cascades that affect cell shape, secretion, integrin activation, metabolic responses, transcriptional responses and cell motility. PARs are 'single-use' receptors: proteolytic activation is irreversible and the cleaved receptors are degraded in lysosomes. Thus, PARs play important roles in 'emergency situations', such as trauma and inflammation. The availability of selective agonists and antagonists of protease inhibitors and of genetic models

[1]*To whom correspondence should be addressed (e-mail: nigelb@itsa.ucsf.edu).*

has generated evidence to suggests that proteases and their receptors play important roles in coagulation, inflammation, pain, healing and protection. Therefore, selective antagonists or agonists of these receptors may be useful therapeutic agents for the treatment of human diseases.

## Introduction

The superfamily of G-protein-coupled receptors (GPCRs) comprises the largest and most functionally diverse groups of signalling molecules. These receptors play essential roles in normal regulation of most biological processes. They are also of great importance in human disease since receptor mutations can cause disease and many of the most commonly prescribed drugs are receptor antagonists or agonists. Although all GPCRs share several structural features, e.g. seven transmembrane domains and conserved motifs, they are able to interact with very diverse agonists, including peptides, lipids, ions and even photons. However, proteases are one of the most intriguing agonists. Certain extracellular proteases, derived from the circulation and inflammatory cells, can specifically cleave and activate protease-activated receptors (PARs), a small, but important, sub-group of the GPCR superfamily. To date, four PARs have been cloned and all share the same basic mechanism of activation: proteases cleave at a specific site within the extracellular N-terminus to expose a new N-terminal tethered ligand domain, which binds to and thereby activates the cleaved receptor (Figure 1).

The concept that a protease cleaves and activates a GPCR raises several questions from the standpoint of receptor regulation, signal transduction and function. What is the molecular mechanism of activation? How can a catalyst, which would ultimately cleave all surface receptors, induce concentration-dependent responses? Since receptor cleavage is an irreversible event, what happens to the cleaved receptors and how do cells recover their ability to respond? Given that PARs are 'one-shot' receptors (once cleaved they can no longer be activated by proteolysis) is it likely that they participate in normal regulation, or would they serve to signal under emergency situations during injury or inflammation? If so, would agonists or antagonists of such receptors serve as therapies for human disease?

This brief article, which focuses on the recent advances in our understanding of the mechanism of activation and function of PARs, attempts to address some of these questions. Several recent articles comprehensively review this field [1–4].

## Cell-surface proteolysis initiates signal transduction

A common theme of signalling by GPCRs is that a single agonist can activate several different receptors, and receptors can respond to many agonists, albeit with varying potencies. Proteases are also capable of activating several distinct PARs. In all cases, the basic mechanism of activation is the same: cleavage at a specific site within the extracellular N-terminus of the receptor exposes a new

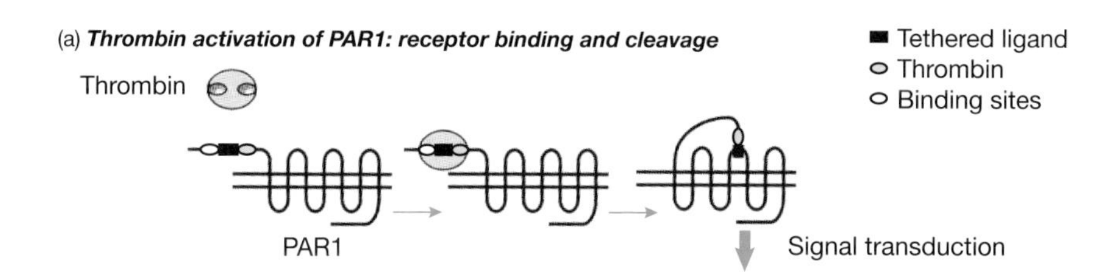


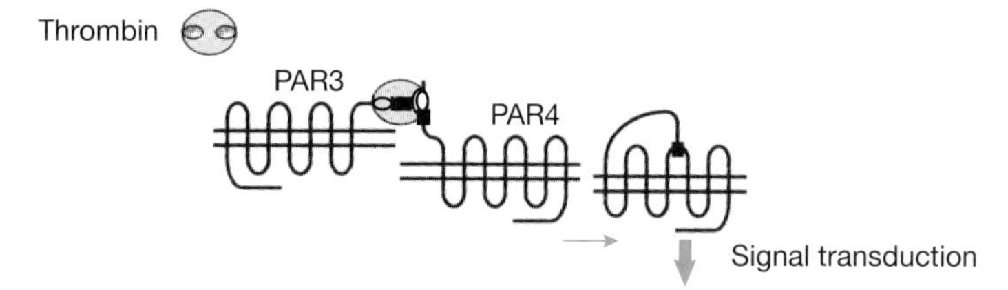


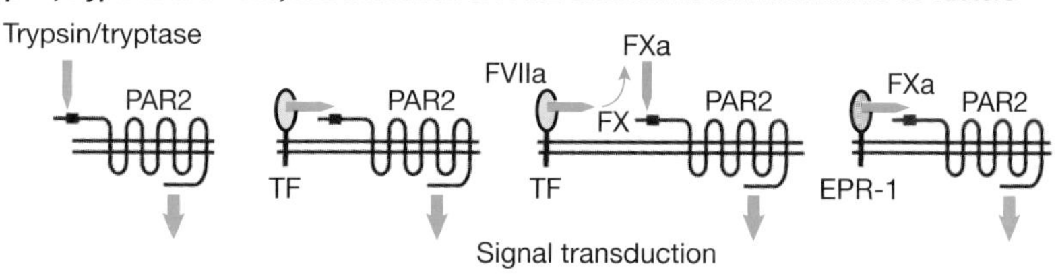


**Figure 1. Mechanisms of activation of PARs**
(a) Thrombin binds to extracellular PAR1 sites and then cleaves them to expose the tethered ligand, which binds and activates the cleaved receptor. (b) Mouse platelets express PAR3 and PAR4. Although murine PAR3 does not signal, it binds thrombin and localizes it close to PAR4 at the cell surface, which facilitates PAR4 cleavage and signalling. (c) Activation of PAR2. Trypsin and tryptase cleave PAR2 without binding. Factor VIIa (FVIIa) can activate PAR2 only if tethered to the cell surface by tissue factor (TF). The FVIIa–TF complex also generates factor Xa (FXa) from factor X (FX) at the cell surface, which facilitates PAR2 activation. EPR-1 may also concentrate FXa at the cell surface to facilitate activation of PAR2. Modified from [1], with permission from *Nature*. ©(2000) Macmillan Magazines Ltd; (www.nature.com).

N-terminus that serves as a tethered ligand by binding to and activating the cleaved receptor. Many factors markedly alter the efficiency of this mechanism. They include the activation of zymogens, the presence of protease inhibitors, the existence of binding sites for proteases on the receptors or at the cell surface, interactions between PARs, and pre- and post-translational modifications of receptors. Thus, the protease agonists of PARs will have different potencies depending upon the cellular environment in which they are acting.

### Thrombin receptors: PAR1, PAR3 and PAR4

Thrombin, which is generated in the circulation during activation of the coagulation cascade, converts fibrinogen into fibrin which leads to clot formation (see Chapter 8). However, thrombin has multiple biological effects, including platelet aggregation and endothelial cell proliferation, both of which are mediated by PARs. Thrombin activates PAR1, PAR3 and PAR4, although the precise mechanisms of activation vary [1] (Figure 1).

Thrombin activates PAR1 in two stages [5] (Figure 1a). First, thrombin binds to PAR1 on either side of the cleavage site. The downstream site has similarities with the C-terminal sequence of hirudin, an anticoagulant thrombin inhibitor from leech saliva. Secondly, thrombin cleaves PAR1 between $Arg^{41}$ and $Ser^{42}$ to expose a new N-terminal tethered ligand domain, SFLLRN. The tethered ligand interacts with domains in extracellular loop 2, which presumably alters the conformation of the receptor to permit coupling to G-proteins. PAR3 also contains the thrombin binding sites and these two receptors have similar $EC_{50}$ values (approx. 0.2 nM), whereas PAR4 lacks thrombin-binding sites and so only responds to higher concentrations of thrombin ($EC_{50}$ approx. 5 nM) [1,6]. PAR4 also responds to trypsin with a similar $EC_{50}$ value.

These differences in the mechanism of activation have interesting functional consequences. For example, human platelets express both PAR1 and PAR4. The existence of two thrombin receptors with different affinities for thrombin allows platelets to respond to graded concentrations of this enzyme in a regulated manner: PAR1 mediates responses to low concentrations of thrombin and PAR4 comes into play at high concentrations. Mouse platelets express PAR3 and PAR4, but not PAR1. Even though mouse PAR3 does contain a binding site for thrombin, mouse PAR3 does not mediate thrombin signalling, even when over-expressed. So how can mouse platelets respond to low concentrations of thrombin if they lack a functional high affinity receptor? The answer lies in an unusual intermolecular interaction between PAR3 and PAR4 (Figure 1b). Thrombin binding to PAR3 transiently anchors the enzyme to the cell surface where it can cleave and activate PAR4.

## Trypsin and tryptase receptor: PAR2

Although trypsin is usually viewed as a digestive enzyme produced in the pancreas, trypsinogens are widely expressed in endothelial and epithelial cells, leucocytes and neurons, and both pancreatic and extra-pancreatic trypsins can signal to many cell types by cleaving and activating PAR2. The trypsin-like serine protease tryptase is expressed by most human mast cells. This enzyme is released as an active, heparin-bound tetramer, which is capable of cleaving neuropeptides, procoagulant proteins and PAR2. Trypsin and tryptase do not bind to PAR2, but rather cleave directly at $Arg^{36}$ and $Ser^{37}$ to expose the tethered ligand SLIGKV [7,8] (Figure 1c). The absence of binding sites may be reflected in their higher $EC_{50}$ values compared with PAR1 and PAR3 activation by thrombin ($EC_{50}$ values: trypsin, approx. 1 nM; tryptase, approx. 100 nM).

The ability of proteases to activate PAR2 can depend on the presence of associated proteins that anchor the proteases to the plasma membrane and thereby concentrate them on the cell surface. High concentrations of coagulation factor VIIa are unable to elicit cellular $Ca^{2+}$ responses [9]. However, in cells that also express tissue factor, an integral membrane protein, factor VIIa is bound and concentrated to the cell surface and cleaves PAR2 ($EC_{50}$ approx. 3.5 nM), and to a much lesser extent PAR1; however, this concentration is much higher than physiological levels. It is intriguing that addition of physiological concentrations of coagulation factor X, which is converted into its active form (factor Xa) by factor VIIa (see Chapter 8), reduces the concentration requirement of factor VIIa ($EC_{50}$ approx. 8 pM) [9]. Coagulation factor Xa can also activate PAR2 by a mechanism independent of factor VIIa and tissue factor. Human vascular endothelial cells express effector cell protease receptor-1 (EPR-1), a high-affinity binding site for factor Xa. EPR-1 binds, and thereby concentrates, factor Xa at the cell surface to facilitate activation of PAR2 ($EC_{50}$ approx. 30 nM) (Figure 1c) [10].

Membrane-bound proteases are perfectly localized to act upon PARs and one such protease has been identified. Membrane-type serine protease 1 is a type II integral membrane protein, which is capable of signalling through PAR2 but not through PAR1, PAR3 or PAR4. Since membrane-type serine protease 1 and PAR2 have similar tissue distribution, it is conceivable that this protease could be a physiological PAR2 agonist.

## Non-mammalian proteases can activate PARs

Recent observations support the intriguing possibility that proteases from parasites and pathogens can 'hijack' mammalian PARs and thereby cause disease. Gingipains are bacterial cysteine proteases with a trypsin-like activity from the bacterium *Porphyromonas gingivalis*. Gingipains cleave after arginine and lysine residues, and can cleave PAR1 and PAR4, thereby causing their activation. This activity has potential implications for certain disease states. Dust mite proteolytic allergens can also activate PAR2, which may contribute to asthma.

## Pre- and post-translational modifications of PARs alter protease signalling

Modifications of PARs can profoundly alter the capacity of proteases to cleave and activate their receptors. A polymorphic form of human PAR2 exists with a point mutation in extracellular loop 2 ($Phe^{204} \rightarrow Ser$) [11]. This variant displays diminished sensitivity to trypsin and SLIGKV, yet enhanced sensitivity to PAR4-derived peptides. It is not known whether there are polymorphic variants of the other PARs and if their expression is associated with human disease.

PARs, like many GPCRs, possess putative glycosylation sites and can be extensively glycosylated, which markedly increases their mass. Recent studies indicate that glycosylation of PAR2 profoundly alters the ability of tryptase to activate this receptor [12]. Tryptase is a 'doughnut-shaped' tetramer with active sites on the inner surface. The extracellular tail of PAR2 contains a glycosylation site in close proximity to the activation site. The potency with which tryptase (but not trypsin) activates PAR2 is dramatically increased by mutation of this sequon, by enzymic deglycosylation of PAR2, or by expres-

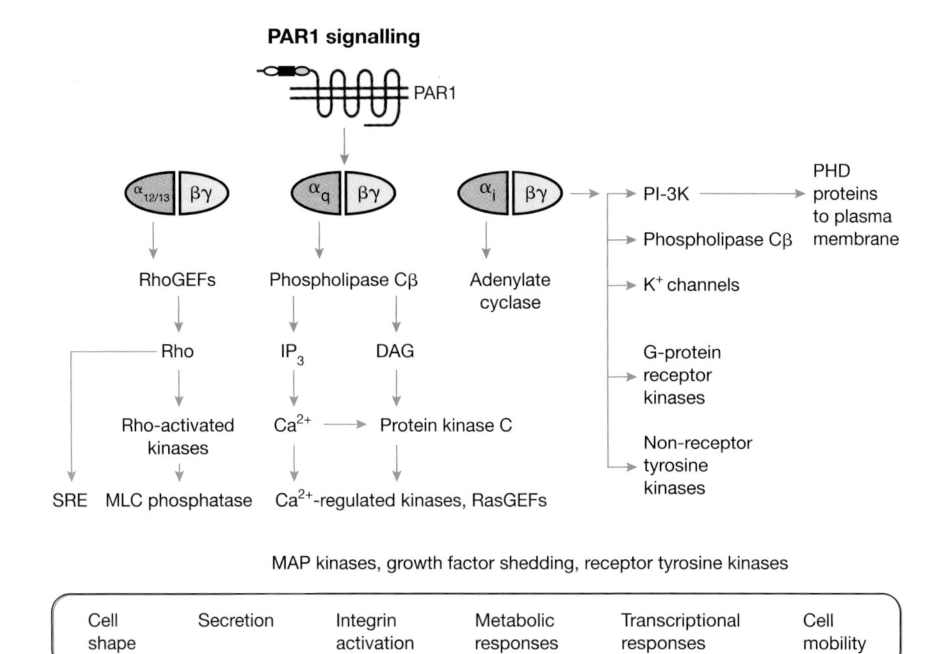


**Figure 2. Summary of the pathways by which PAR1 induces alterations in cell shape, secretion, integrin activation, metabolism, transcription and cell motility**
PAR1 couples to several different G-proteins. The α-subunit of $G_{12}$ and $G_{13}$ binds to RhoGEFs (guanine nucleotide exchange factors that activate small G-proteins such as Rho) to provide a pathway for altering cell shape and migration. $G\alpha_q$ couples to pathways that trigger MAP kinases, which regulate transcriptional responses, integrin activation, granule secretion, and metabolic responses. Gα1 inhibits adenylate cyclase. Gβγ subunits activate phosphoinositide 3-kinase, which itself couples to multiple pathways. SRE, serum response element; PI-3K, phosphoinositide 3-kinase; $IP_3$, inositol 1,4,5-trisphosphate; DAG, diacylglycerol; PHD, pleckstrin homology domain; MLC, myosin light-chain. Modified from [1], with permission from *Nature*. ©(2000) Macmillan Magazines Ltd; (www.nature.com).

sion of PAR2 in glycosylation-defective cells, such that tryptase becomes almost as potent as trypsin. Perhaps the deglycosylated receptor is more readily accommodated by the active site of tryptase, allowing for efficient cleavage.

## Initiation and termination of signal transduction

Once activated, PARs can couple to several heterotrimeric G-proteins and thereby trigger a cascade of signalling events that result in marked phenotypic changes. PAR1 can couple to $G_{12/13}$, $G_q$ and $G_i$ families, whereas PAR2 couples to $G_q$. This coupling activates signalling pathways that alter cell motility, secretion, shape, growth and survival (Figure 2). Low concentrations of a protease could eventually cleave all receptors on the surface of a cell. However, the enzyme concentration determines the rate of receptor activation and thus the rate of second messenger generation, which permits concentration-dependent responses. This dose-response is achieved because

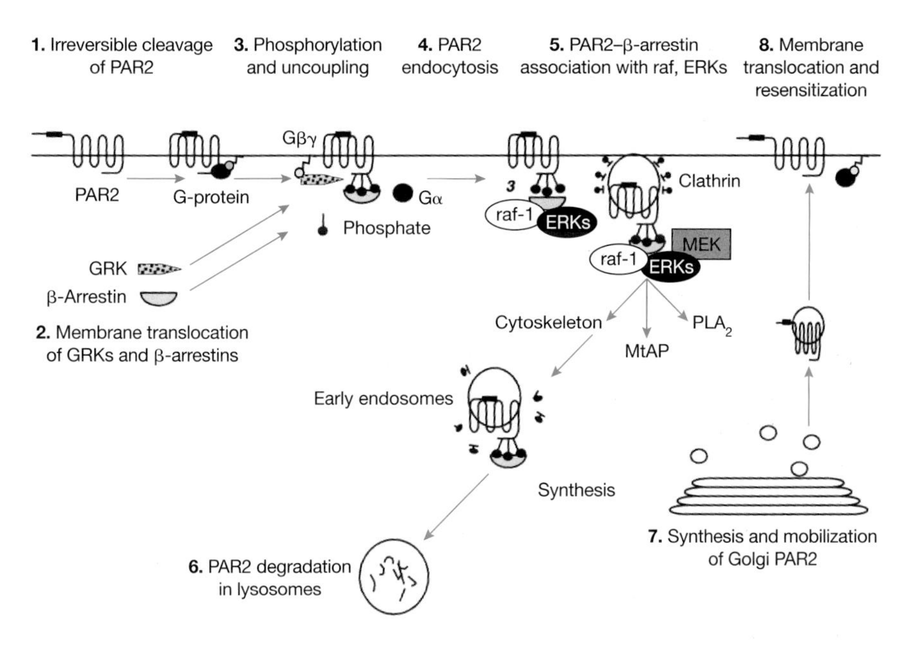


**Figure 3. Mechanism and function of agonist-induced trafficking of PAR2**
(1) PAR2 agonists cleave the receptor to induce (2) membrane translocation of G-protein receptor kinases (GRKs) and β-arrestins. (3) β-Arrestins interact with GRK-phosphorylated PAR2 to uncouple the receptor from G-proteins and to terminate the signal; they are also adaptors for endocytosis (4) at sites of clathrin-coated pits, which pinch off from the plasma membrane in a dynamin-dependent process. (5) β-Arrestin forms a complex with PAR2, raf-1 and activated ERK1 and ERK2 (extracellular signal regulated kinases). Cytosolic activated ERK1 and ERK2 phosphorylate cytoskeletal proteins, microtubule associated proteins (MtAPs), and phospholipase $A_2$ ($PLA_2$). (6) Endocytosed PAR2 is targeted to lysosomes. (7 and 8) Resensitization requires mobilization of PAR2 from Golgi stores or synthesis of new receptors. MEK, mitogen-activated protein kinase kinase.

the activated receptors are quickly shut off, terminating their signalling after the production of a 'quantum' or 'packet' of second messenger [13]. Thus, the level of second messenger (or response) generated is proportional to the number of receptors activated in a given time period and not to the total number of receptors cleaved by the protease agonist. Even though a low concentration of enzyme could eventually cleave all surface receptors, it is the rate at which they are cleaved that determines the cellular response to the enzyme concentration.

After proteolytic activation, the tethered ligand of a PAR is always available to interact with the cleaved receptor. However, PAR signalling is rapidly terminated by mechanisms similar to those used by other GPCRs, namely uncoupling from heterotrimeric G-proteins. PAR activation triggers receptor phosphorylation by G-protein receptor kinases and second messenger kinases. β-Arrestins translocate from the cytosol to the plasma membrane where they interact with phosphorylated receptors to mediate uncoupling and desensitization [14] (Figure 3).

Activated PARs internalize into early endosomes at sites of clathrin-coated pits. β-Arrestins couple PAR2 to clathrin, and are thus required for agonist-induced receptor endocytosis. The GTPase dynamin is required for the final stages of endosome formation. Internalized PARs are mostly destined for degradation in lysosomes. Therefore, resensitization of responses to proteases requires either the mobilization of preformed pools of receptors or the synthesis of new receptors (Figure 3).

In addition to mediating PAR uncoupling and endocytosis, β-arrestins also play an important role in signal transduction (Figure 3). β-Arrestin-dependent endocytosis of PAR2 is required for activation of the mitogen-activated protein (MAP) kinase cascade [15]. β-Arrestins serve as molecular scaffolds that recruit and organize various upstream components of the MAP kinase pathway (e.g. raf-1) into endosomes. This process determines the sub-cellular location of activated MAP kinases and thereby governs their specificity and function.

Cell surface proteolysis also contributes to terminating PAR signalling. Certain proteases cleave PARs to remove or destroy the tethered ligand domain, thereby rendering them unresponsive to protease agonists. For instance, neutrophil cathepsin G cleaves PAR1 and PAR3 to form receptors that are unresponsive to thrombin. In a similar manner, mast cell chymase inactivates PAR1 and PAR2.

## Role of PARs in health and disease states

The appreciation that proteases can serve as signalling molecules has provided new insights into the physiological and pathophysiological functions of these enzymes (Figure 4). However, an understanding of the functions of PARs has been hampered by several obstacles. The first obstacle is the lack of highly selective and potent agonists. Proteases are not absolutely selective for one PAR. Proteases such as thrombin and trypsin can cleave several PARs, as well as other proteins, which may account for their biological actions. Synthetic peptides that correspond to the tethered ligand domains of PAR1, PAR2 and PAR4 can directly activate their receptor and are useful pharmacological tools. However, there is some cross-reactivity (for example, the PAR1 tethered ligand peptide SFLLRN can also activate PAR2). Moreover, these activating peptides are weak agonists that are effective only at micromolar concentrations, where some peptides can have non-specific effects. The second obstacle is that, with the exception of PAR1, there are no selective antagonists of PARs. The use of protease inhibitors is one strategy for understanding the functions of proteases; however, a more selective approach has been the use of genetically modified mice that lack or overexpress PARs.

PARs are disposable ‘one-shot’ receptors, as they are activated in an irreversible manner and then degraded. Given this seemingly wasteful mechanism of activation it is unlikely that PARs mediate routine intercellular signalling. Proteases and PARs instead appear to play important roles in ‘emergency

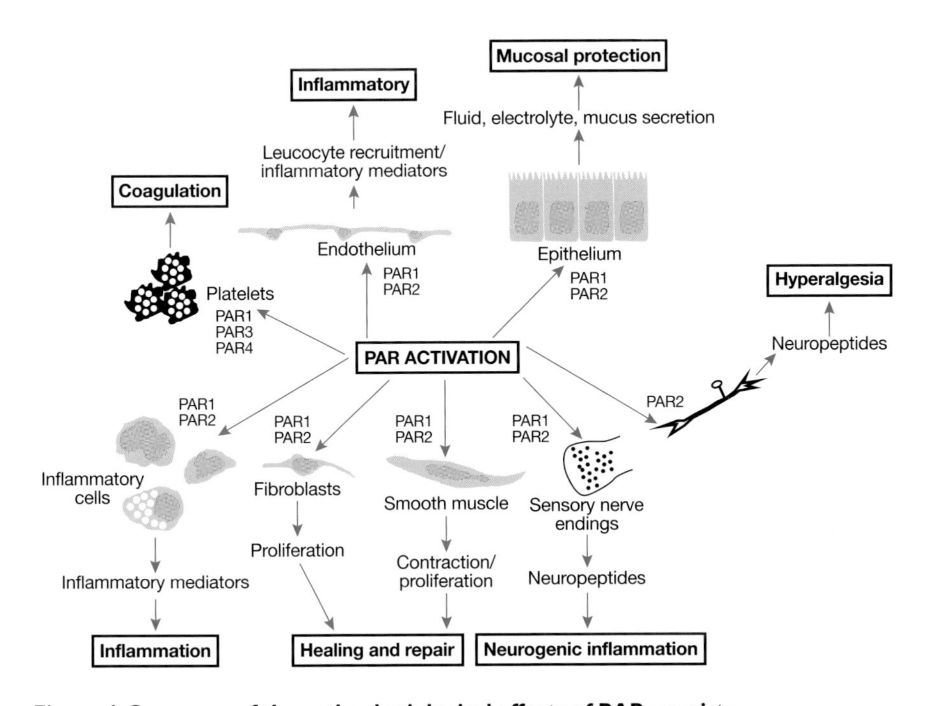


**Figure 4. Summary of the pathophysiological effects of PAR agonists**
PAR1 and PAR2 are expressed by endothelial cells, epithelial cells, neurons, myocytes, fibroblasts and inflammatory cells. PARs trigger pathways related to inflammation, hyperalgesia, tissue repair and protection in these cell types. Modified from Vergnolle, N., Wallace, J.L., Bunnett, N.W. & Hollenberg, M.D. (2001) Protease-activated receptors in inflammation, neuronal signaling and pain. *Trends Pharmacol. Sci.* **22**, 146–152, with permission from Elsevier Science.

situations', e.g. during coagulation or when mast cells degranulate. Indeed, accumulating evidence suggests that PARs play roles in inflammation, hyperalgesia, tissue repair and cancer biology.

## Pro-inflammatory effects

Agonists of PARs can trigger all of the hallmarks of inflammation: swelling, redness, heat, pain and tissue repair (Figure 4). Thus, thrombin activates PAR1 on endothelial cells to induce vasodilatation, elevated vascular permeability to plasma proteins and increased rolling and adhesion of leucocytes. PAR1 agonists also activate inflammatory cells (mast cells, lymphocytes, neutrophils) to trigger release of chemoattractive agents and inflammatory mediators (e.g. histamine, cytokines, eicosanoids). The availability of thrombin inhibitors, PAR1 antagonists and PAR1$^{-/-}$ mice has facilitated recent investigations into the role of PAR1 in animal models of disease. Renal inflammation is markedly diminished by the thrombin inhibitor hirudin and is attenuated in PAR1$^{-/-}$ mice. Hirudin also ameliorates collagen-induced arthritis by inhibiting thrombin.

PAR2 also plays an important pro-inflammatory role. PAR2 is upregulated on endothelial cells by pro-inflammatory agents, and PAR2 agonists, such as tryptase from mast cells, cause increased vascular permeability, systemic hypotension and bronchoconstriction. PAR2 agonists also induce the rolling, adherence and recruitment of leucocytes in venules by a mechanism that is dependent on platelet-activating factor release. Observations in PAR2 knockout mice suggest a role for this receptor in leucocyte adhesion to venules after surgery.

## Neurogenic inflammation

Some of the pro-inflammatory effects of PAR1 and PAR2 agonists are mediated by neurogenic mechanisms (Figure 5). Neurogenic inflammation is a form of inflammation that is controlled by the sensory nervous system and that depends on the release of the neuropeptides substance P (SP) and calcitonin gene related peptide (CGRP) from the peripheral endings of

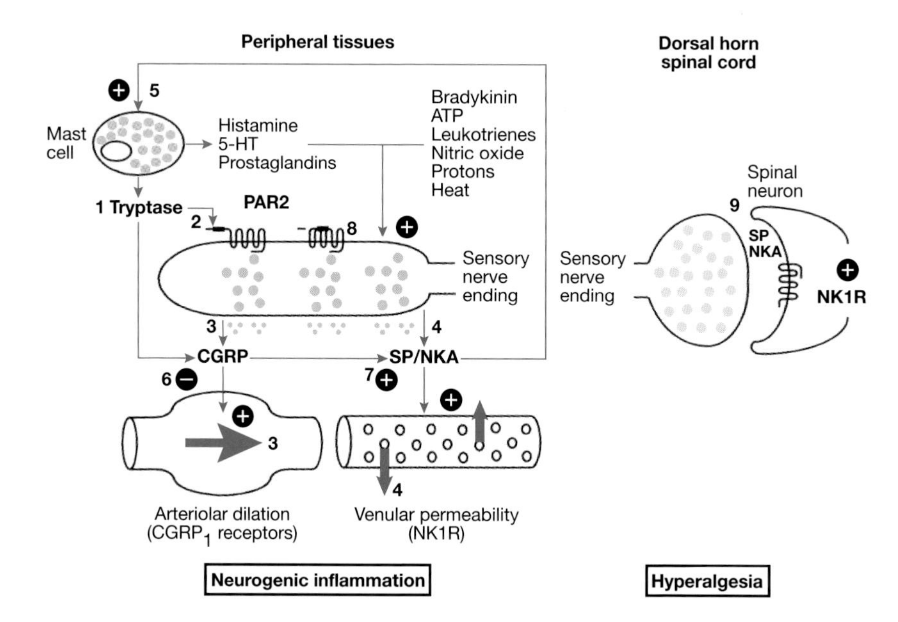


**Figure 5. Neurogenic mechanisms of PAR2-induced oedema and hyperalgesia**
(1) Tryptase released from degranulated mast cells cleaves PAR2 at the plasma membrane of sensory nerve endings to expose a tethered ligand domain that binds to and activates the cleaved receptor. (2) Activation of PAR2 stimulates the release of CGRP and the tachykinins, SP and neurokinin A (NKA) from sensory nerve endings. (3) CGRP interacts with the $CGRP_1$ receptor to induce arteriolar dilatation and hyperaemia. (4) SP interacts with the NK1R on endothelial cells of post-capillary venules to cause gap formation and plasma extravasation. The hyperaemia and plasma extravasation cause oedema. (5) SP may stimulate the degranulation of mast cells, thereby providing positive feedback. (6) Tryptase degrades CGRP and terminates its effects. (7) CGRP inhibits SP degradation by neutral endopeptidase and also enhances SP release, thereby amplifying the effects. (8) Mediators from mast cells and other inflammatory cells stimulate the release of vasoactive peptides from sensory nerves; they also sensitize nerves. (9) Sub-inflammatory doses of PAR2 agonists also induce central sensitization and both thermal and mechanical hyperalgesia, which is likely to depend on enhanced release of SP and activation of the NK1R on spinal neurons. 5-HT, 5-hydroxytryptamine. Modified from Steinhoff, M., Vergnolle, N., Young, S.H., Tognetto, M., Amadesi, S., Ennes, H.S., Trevisani, M., Hollenberg, M.D., Wallace, J.L., Caughey, G.H. et al. (2000) Agonists of proteinase-activated receptor 2 induce inflammation by a neurogenic mechanism. *Nat. Med.* **6**, 151–158, with permission from Nature Publishing Group.

primary spinal afferent neurons. Certain inflammatory agents trigger the release of these peptides within peripheral tissues such as the skin, airway and intestine, where they interact with neurokinin$_1$ receptors (NK1Rs) and CGRP$_1$ receptors to cause hyperaemia, plasma extravasation and recruitment of granulocytes. Recent studies suggest that thrombin from the circulation and tryptase from mast cells, which are in close association with sensory nerves, can signal to sensory nerves through PAR1 and PAR2 [16]. A large proportion of small diameter neurons in the dorsal root ganglia, which contain SP and CGRP, also express PAR1 and PAR2. Thrombin and tryptase can directly signal to these neurons to release SP and CGRP. When injected into rat paws, PAR1 and PAR2 agonists cause a severe oedema that lasts for several hours and that is accompanied by an intense granulocyte infiltration. Both PAR1- and PAR2-induced oedema is inhibited by NK1R and CGRP$_1$ receptor antagonists or by ablating C-fibres with capsaicin. Thrombin and tryptase can similarly signal to the enteric nervous system, which may have marked effects on gastrointestinal secretion and motility.

## Hyperalgesia

The discovery of receptors for proteases on nociceptive sensory neurons raised the possibility that activation of PAR1 or PAR2 on these neurons would lead to central transmission of a signal, and particularly nociceptive messages. A recent study points to a direct role of proteases and their receptors in somatic pain [17]. Injections into the paw of sub-inflammatory doses of PAR2 agonists in rats and mice induce a prolonged and sustained thermal and mechanical hyperalgesia. Hyperalgesia is not observed in PAR2$^{-/-}$ animals, confirming involvement of this receptor. Additionally, deletion of the NK1R and preprotachykinin A, which encodes SP and neurokinin A, as well as the central administration of NK1R antagonists and cyclo-oxygenase inhibitors attenuates the hyperalgesia. Together, these results suggest that the hyperalgesia depends on the central activation of NK1R and the release of prostaglandins within the spinal cord (Figure 5). Remarkably, the hyperalgesia that follows intraplantar administration of formalin and of compound 48/80, which degranulates mast cells, is also diminished in PAR2$^{-/-}$ mice, suggesting an important role for this receptor in pain transmission. The involvement PAR1, PAR3 and PAR4 in nociception remains to be determined.

## Protection

In addition to their pro-inflammatory effects, there is evidence in some systems that PAR agonists can have protective roles. Trypsin in the intestinal lumen can signal to enterocytes by cleaving apical PAR2 to release prostaglandins, which have protective functions in the gastrointestinal tract. Agonists of PAR2 can trigger the relaxation of murine airway through the release of prostaglandin E$_2$, thereby inducing a strong bronchodilatation, and can diminish infiltration of neutrophils in response to bacterial lipopolysaccharide [18,19]. After myocardial ischaemia, infusion of PAR2 agonists significantly improves cardiac function and reduces tissue damage [20]. Some of these protective effects may be mediated by a neurogenic mechanism, and there is evidence that sensory nerves and CGRP have protective functions in several tissues. For example, PAR2 agonists induce secretion of mucus by triggering the release of CGRP, and thereby protect the gastric mucosa from experimentally induced ulceration [21].

### Tissue repair and cell proliferation

Agonists of PAR1 and PAR2 can stimulate proliferation of several cell types, including endothelial cells, myocytes and fibroblasts, which suggests a role for PARs in repair, angiogenesis and wound healing. A role for PAR1 in vascular injury is suggested by the observation that a potent and selective PAR1 antagonist RWJ-58259 markedly reduces neointimal thickness in a model of vascular restenosis induced by balloon injury in rats [22]. Observations in PAR1$^{-/-}$ mice also suggest a role for this receptor in the regulation of extracellular matrix formation and remodelling associated with vascular injury. A role for PAR1 in embryonic development is indicated by the finding that half of PAR1 knockout mice die *in utero*, a defect rescued by targeted overexpression of PAR1 in endothelial cells. Thus, PAR1 may play a role in normal vascular development [23].

Agonists of PARs have been found to have roles in cell proliferation, tumour cell invasion and metastasis. There appears to be a correlation between PAR1 expression and metastatic potential of tumour cells [24]. For example, PAR1 antisense oligonucleotides, which reduce PAR1 expression, also diminish the invasive nature of tumour cells. The molecular mechanism of the PAR1 involvement is thought to involve integrin recruitment to focal adhesion sites, thereby allowing migration of the cells.

There is increasing evidence that trypsin may play a major role in proliferation and cancer progression. Trypsin is produced by many lung, colon and ovarian tumours and also by endothelial cells in close proximity to gastric tumours. Activation of PAR2 by trypsin and the peptide ligand (SLIGKV) induces proliferation of human gastric and colon cancer cell lines.

## Conclusions

To date, four PARs have been cloned, and there is accumulating pharmacological evidence for the existence of new receptors or new subtypes of existing receptors. The number of proteases known to cleave and activate receptors is growing rapidly. Considerable progress has been made in characterizing the molecular mechanisms of receptor activation and signal transduction. The availability of selective agonists and antagonists, of protease inhibitors and of genetic models has generated evidence to suggests that proteases and their receptors play important roles in coagulation, inflammation, pain, healing and protection.

There is, however, much still to learn. Many of the biological functions of proteases cannot be accounted for by the known PARs, which suggests the existence of additional mechanisms. In many systems, the physiologically important agonists of PARs are unknown. The tools to study proteases and PARs in animal models of disease have not been applied to many systems, and there are no selective antagonists of PAR2, PAR3 and PAR4. Finally, although the experimental evidence in animal models is tantalizing, little is known about the direct role of PARs in human disease.

## Summary

- *Certain proteases from the circulation and from inflammatory cells can directly signal to cells by cleaving PARs, members of a new sub-family of heptahelical receptors that couple to G-proteins.*
- *Cleavage exposes a tethered ligand domain that binds to and activates the cleaved receptors. Activated PARs couple to signalling cascades that affect cell shape, secretion, integrin activation, metabolic responses, transcriptional responses and cell motility.*
- *PARs are 'single use' receptors: proteolytic activation is irreversible and the cleaved receptors are degraded in lysosomes. Thus, PARs play important roles in 'emergency situations' such as trauma and inflammation.*
- *Proteases and their receptors play important roles in coagulation, inflammation, pain, healing and protection.*
- *Selective antagonists or agonists of these receptors may be useful therapeutic agents for the treatment of human diseases.*

Research in the authors' laboratory is funded by National Institutes of Health grants DK39957, DK43207, DK57840 and a Focussed Giving Grant from the R.W. Johnson Foundation.

## References

1. Coughlin, S.R. (2000) Thrombin signalling and protease-activated receptors. *Nature (London)* **407**, 258–264
2. Macfarlane, S.R., Seatter, M.J., Kanke, T., Hunter, G.D. & Plevin, R. (2001) Proteinase-activated receptors. *Pharmacol. Rev.* **53**, 245–282
3. Vergnolle, N. (2000) Proteinase-activated receptors – novel signals for gastrointestinal pathophysiology. *Aliment. Pharmacol. Ther.* **14**, 257–266
4. Cocks, T.M. & Moffatt, J.D. (2000) Protease-activated receptors: sentries for inflammation? *Trends Pharmacol. Sci.* **21**, 103–108
5. Vu, T.K., Hung, D.T., Wheaton, V.I. & Coughlin, S.R. (1991) Molecular cloning of a functional thrombin receptor reveals a novel proteolytic mechanism of receptor activation. *Cell* **64**, 1057–1068

6. Xu, W., Andersen, H., Whitmore, T.E., Presnell, S.R., Yee, D.P., Ching, A., Gilbert, T., Davie, E.W. & Foster, D.C. (1998) Cloning and characterization of human protease-activated receptor 4. *Proc. Natl. Acad. Sci. U.S.A.* **95**, 6642–6646
7. Nystedt, S., Emilsson, K., Wahlestedt, C. & Sundelin, J. (1994) Molecular cloning of a potential proteinase activated receptor. *Proc. Natl. Acad. Sci. U.S.A.* **91**, 9208–9212
8. Molino, M., Barnathan, E.S., Numerof, R., Clark, J., Dreyer, M., Cumashi, A., Hoxie, J., Schechter, N., Woolkalis, M. & Brass, L.F. (1997) Interactions of mast cell tryptase with thrombin receptors and PAR-2. *J. Biol. Chem.* **272**, 4043–4049
9. Camerer, E., Huang, W. & Coughlin, S.R. (2000) Tissue factor- and factor X-dependent activation of protease-activated receptor 2 by factor VIIa. *Proc. Natl. Acad. Sci. U.S.A.* **97**, 5255–5260
10. Bono, F., Schaeffer, P., Herault, J.P., Michaux, C., Nestor, A.L., Guillemot, J.C. & Herbert, J.M. (2000) Factor Xa activates endothelial cells by a receptor cascade between EPR-1 and PAR-2. *Arterioscler. Thromb. Vasc. Biol.* **20**, E107–E112
11. Compton, S.J., Cairns, J.A., Palmer, K.J., Al-Ani, B., Hollenberg, M.D. & Walls, A.F. (2000) A polymorphic protease-activated receptor 2 (PAR2) displaying reduced sensitivity to trypsin and differential responses to PAR agonists. *J. Biol. Chem.* **275**, 39207–39212
12. Compton, S.J., Renaux, B., Wijesuriya, S.J. & Hollenberg, M.D. (2001) Glycosylation and the activation of proteinase-activated receptor 2 (PAR(2)) by human mast cell tryptase. *Br. J. Pharmacol.* **134**, 705–718
13. Ishii, K., Hein, L., Kobilka, B. & Coughlin, S.R. (1993) Kinetics of thrombin receptor cleavage on intact cells. Relation to signaling. *J. Biol. Chem.* **268**, 9780–9786
14. Dery, O., Thoma, M.S., Wong, H., Grady, E.F. & Bunnett, N.W. (1999) Trafficking of proteinase-activated receptor-2 and beta-arrestin-1 tagged with green fluorescent protein. beta-Arrestin-dependent endocytosis of a proteinase receptor. *J. Biol. Chem.* **274**, 18524–18535
15. DeFea, K.A., Zalevsky, J., Thoma, M.S., Dery, O., Mullins, R.D. & Bunnett, N.W. (2000) Beta-arrestin-dependent endocytosis of proteinase-activated receptor 2 is required for intracellular targeting of activated ERK1/2. *J. Cell Biol.* **148**, 1267–1281
16. Steinhoff, M., Vergnolle, N., Young, S.H., Tognetto, M., Amadesi, S., Ennes, H.S., Trevisani, M., Hollenberg, M.D., Wallace, J.L., Caughey, G.H. et al. (2000) Agonists of proteinase-activated receptor 2 induce inflammation by a neurogenic mechanism. *Nat. Med.* **6**, 151–158
17. Vergnolle, N., Bunnett, N.W., Sharkey, K.A., Brussee, V., Compton, S.J., Grady, E.F., Cirino, G., Gerard, N., Basbaum, A.I., Andrade-Gordon, P. et al. (2001) Proteinase-activated receptor-2 and hyperalgesia: a novel pain pathway. *Nat. Med.* **7**, 821–826
18. Cocks, T.M., Fong, B., Chow, J.M., Anderson, G.P., Frauman, A.G., Goldie, R.G., Henry, P.J., Carr, M.J., Hamilton, J.R. & Moffatt, J.D. (1999) A protective role for protease-activated receptors in the airways. *Nature (London)* **398**, 156–160
19. Cocks, T.M. & Moffatt, J.D. (2001) Protease-activated receptor-2 (PAR2) in the airways. *Pulm. Pharmacol. Ther.* **14**, 183–191
20. Napoli, C., Cicala, C., Wallace, J.L., de Nigris, F., Santagada, V., Caliendo, G., Franconi, F., Ignarro, L.J. & Cirino, G. (2000) Protease-activated receptor-2 modulates myocardial ischemia-reperfusion injury in the rat heart. *Proc. Natl. Acad. Sci. U.S.A.* **97**, 3678–3683
21. Kawabata, A., Kinoshita, M., Nishikawa, H., Kuroda, R., Nishida, M., Araki, H., Arizono, N., Oda, Y. & Kakehi, K. (2001) The protease-activated receptor-2 agonist induces gastric mucus secretion and mucosal cytoprotection. *J. Clin. Invest.* **107**, 1443–1450
22. Andrade-Gordon, P., Derian, C.K., Maryanoff, B.E., Zhang, H.C., Addo, M.F., Cheung, W., Damiano, B.P., D'Andrea, M.R., Darrow, A.L., de Garavilla, L. et al. (2001) Administration of a potent antagonist of protease-activated receptor-1 (PAR-1) attenuates vascular restenosis following balloon angioplasty in rats. *J. Pharmacol. Exp. Ther.* **298**, 34–42
23. Griffin, C.T., Srinivasan, Y., Zheng, Y.W., Huang, W. & Coughlin, S.R. (2001) A role for thrombin receptor signaling in endothelial cells during embryonic development. *Science* **293**, 1666–1670

24. Even-Ram, S., Uziely, B., Cohen, P., Grisaru-Granovsky, S., Maoz, M., Ginzburg, Y., Reich, R., Vlodavsky, I. & Bar-Shavit, R. (1998) Thrombin receptor overexpression in malignant and physiological invasion processes. *Nat. Med.* **4**, 909–914
25. Vergnolle, N., Wallace, J.L., Bunnett, N.W. & Hollenberg, M.D. (2001) Protease-activated receptors in inflammation, neuronal signaling and pain. *Trends Pharmacol. Sci.* **22**, 146–152

# 14

# Mining proteases in the genome databases

David Coates[1]

*School of Biology, University of Leeds, Leeds LS2 9JT, U.K.*

## Abstract

Protease data mining can take advantage both of the many specialist, Web-available databases that cover the genetic, protein and nucleic acid sequence information that is specific to a variety of organisms, and of a flexible, but defined, classification system. However, precomputed data, such as gene predictions, should be used with care. Unless there is definitive supporting information, ideally sequencing of a cDNA to show that the predictions are accurate, followed by expression and biochemical characterization of the predicted protein, the predicted gene and its product remains a possibility, rather than a certainty.

## Introduction

The publication of whole genome sequences in recent years has completely changed the way we look at small gene families, as exemplified by several of the protease genes. Proteases lend themselves to detailed analysis, in part because they have defined active sites that can be easily identified from raw sequence data. While analysis of a genome sequence once left the researcher in a state of wonder, it is now seen as a routine procedure, with the concomitant danger that, having done the sequencing, the rest is seen as merely a cleaning up job. This is not just an oversimplification, but is totally wrong. The enormous change in the rate at which whole genome sequence can be generated has come about because of technical development in sequencing technologies and

[1]*E-mail: d.coates@leeds.ac.uk*

computer power, just as much as because of changed views of the best way to go about it, whether that be step-by-step, or whole genome shotgunning.

## Genome sequencing strategies

The genome sequence of the nematode *Caenorhabditis elegans* was produced over a long period by the 'brick-by-brick' approach [1]. At the same time as the sequence was being generated, the computational tools to analyse it were also in development: tools that could look at a genomic sequence and attempt to predict gene structure from this, so that other algorithms already in development could be used in comparative analyses, to try and identify the potential nature of the 'new' gene(s). When this analysis first started, personal computers were just starting to make an impact, the Web was young and 'real' bioinformatics research was carried out using Unix or Vax/VMS mainframe systems.

By contrast, the genome of *Drosophila melanogaster* was sequenced using a combination of mapping data and shotgun sequencing [2], which produced a mass of sequence data that were, in the first instance, constructed using computational algorithms. This process involves putting together sets of contiguous overlapping clones by sequence analysis, to form chunks of aligned sequence, referred to as 'contigs'. These were checked using a scaffold made in the 'traditional' way. A physical map, based on yeast artificial chromosome and bacterial artificial chromosome clones (both are means of cloning very large fragments of DNA) whose ends had been sequenced (the 'scaffold'), were mapped onto the genome sequence, to confirm the order of the contigs. Whole genome shotgun sequencing had been seen as ideal for sequencing 'simple' genomes (where simple means 'very little repeated sequence', rather than unsophisticated), but was thought not to be suitable for the complex genomes of higher eukaryotes, which are characterized by large amounts of repeated sequences. At the moment, as one might expect, a combination of the two methods is proving to be the best approach.

## Identification of protease genes and gene products

Before the availability of complete genome sequences, protease families were identified because sequence homologues of an identified enzyme were found, or aberrant phenotypes were linked to novel enzymes. Now, relatively simple computer-based searches can identify whole groups of novel protease-like gene products [3]. This can cause significant problems: previously, there would be other information relevant to the protease — enzyme activity or aberrant activity, a phenotype or, at the very least, proof of expression (usually cDNA expression, with the cDNA often being from a specific developmental stage or tissue). A prediction from genome sequence gives virtually no other information, but the neatness of a predicted gene structure can seduce the investigator into believing not only that it is right, but that it is real.

## Gene prediction and genome annotation

Obtaining the sequence is only the first step in trying to identify potential genes of interest. As before, the comparison between the nematode and fruit fly approaches to gene and genome annotation can be instructive by showing how the technology has progressed and become more subtle. Once a reliable sequence is available, the key steps are identifying introns and exons by looking for matched donor and acceptor splice sites within an open reading frame and then trying to join these together to produce something with transcriptional start and end sites [4]. As one might expect, predicting internal exons is much easier than start/stop or non-translated exons. Annotating a genome is done in two parts. The first part is an automatic, algorithm-based approach, where it is the computer that predicts the structures within the genome. In the second phase, a human annotator looks at the output and determines whether or not it looks like a 'good' prediction, at which point it is published. Clearly this could be a serious bottleneck, and highlights the two different approaches to releasing data for everyone else to work on: the 'curated' approach [i.e. SwissProt; see www.ebi.ac.uk/swissprot/Information/information.html (January 2002)] and the automatic release [i.e. GenBank; see www.ncbi.nlm.nih.gov/Genbank/index.html (January 2002)]. It is important to know what kind of data you are looking at, as curated data may be more robust, but have an inbuilt human opinion, whereas directly released, automatically generated data have a bias built in from the algorithms used to generate it, and errors that a human annotator might have prevented. One aspect of this is the assumption, sometimes made by the annotators, that sequences with high sequence similarity between species are orthologous, i.e. they arose in the speciation process from the same gene, and therefore have the same or very similar function. In particular, substrate specificities of highly similar peptidases may be very different from each other, being driven by varying selection on the different organisms.

Nematode gene sequences were originally based on predictions using GeneFinder (P. Green, unpublished work). As one would expect, the predictions were frequently wrong in the details (the exact combination of exons, and the start and end exons in particular), while being correct in the generality. This is an important point because, in the vast majority of cases, it is not a specific short sequence that identifies the gene predictions of interest, but a more general sequence similarity at the predicted protein level. Over the period of the sequencing projects, gene prediction programs have improved, and different approaches have been applied — in bioinformatics terms, the most significant step forward has been the development of hidden Markov models (HMMs; http://hmmer.wustl.edu) [5], for use in predictions as well as in identification of members of gene families. Sequence comparison programs, such as BLAST [6] and FASTA [7], depend on comparing all the bits of a query sequence with all the bits of sequence in the database. The results of such a search are clearly heavily biased to the query sequence — obviously, the search will find close sequence relatives, rather than distant ones. A reiterative process, using the results of a search to trigger a new search, can broaden the result set (as occurs with PsiBlast [8]), as can an 'all-against-all' clustering method such as was used in the *D. melanogaster* annotation exercise {see www.fruitfly.org/annot/similarity.html (January 2002) and [9]}, but it is still a rather crude approach. It also assumes that all members of a group will be built along the same lines. It is, for instance, unlikely that all the sequence segments in a protein would be maintained at the same level of similarity — rather that functionally more relevant parts stay more similar, and other parts change within broader constraints. This may be particularly true of the substrate specificity sites, which might be expected to vary between copies in the same genome, while the catalytic residues are conserved, because the different paralogues are carrying out the same function on different substrates. This approach was developed for the identification of members of protein families by generating HMMs for particular types of protein and then testing predicted gene sets with the relevant HMM to look for new members of the family. The use of an experimentally tested dataset is key to the process of generating an HMM. Protein sequences of known function, essentially functional domains, are aligned and checked, especially to ensure that catalytically or functionally active residues are properly aligned. This alignment is then used to generate a form of consensus, with a probability for each kind of changed amino acid at each position. Further advances have led to the development of neural network systems, which use learning approaches to define the key elements of a functional domain, but the key point is that the searches are based on verified functional data, without introducing a human bias in the determination of what residues are important.

It is also worth remembering that techniques that rely on sequence similarity over large sections of the predicted protein sequence will identify all members of a protease family, including those with active sites that are predicted to be non-functional. An important observation from both the *D. melanogaster* and *C. elegans* genomes, which have been fully sequenced, is the number of metalloproteases that are predicted to have no catalytic activity, on the basis that they lack critical residues. This is not to say that such genes are non-functional; there are many gene products, such as those involved in the complement cascade, that are classed as 'protease-like' by sequence similarity, that are not proteolytic enzymes but that do have other important functions.

These developments have gradually changed the quality of the predictions made on the *C. elegans* data, such that later predictions are much more robust, and it is now rare to find 'simple' errors such as a lack or excess of internal exons. By the time the *D. melanogaster* genome was ready for annotation in December 1999, it was possible to bring together all of these developments.

There is no right way or wrong way to work with sequence data. The principle accepted for *D. melanogaster* annotation was as follows: analyse with every algorithm available; collect all the data together; and then ask a human to make the call as to whether an automatic annotation is considered 'real'. The need for a final human check has led to a variety of interfaces for the annotator or the interested investigator to be able to look at the evidence, and these usually involve a diagrammatic representation of the sequence and any other relevant information. The 'any other relevant information' is the result of the pre-analysis, where predicted coding sequences have been tested against a collection of available databases, such as unigene sets from other sequenced genomes (including both cDNA and protein comparisons) and a range of motif searching programs [4].

## Sources of data on proteases

Proteases lend themselves to genomic analysis in part because many of them have very distinct active sites, with highly conserved sequences of amino acid residues, whose coding sequence can be relatively easily searched for in genome sequence data. These attributes are what have been used to generate the predictions about genes and gene families that populate the sequence annotations. Using the variety of comparative methods detailed above, these data can be brought together visually using, for instance, the InterPro system from EBI (see [10] and www.ebi.ac.uk/interpro/). Annotated predictions are then made available through genome databases and specialist web sites. Organizing this large collection of data is itself a problem, now being addressed by initiatives such as the Ensembl ([11] and www.ensembl.org/), and Gene Ontology ([12] and www.geneontology.org) projects. One of the key sources, and the best starting point, for getting information on protease genes and gene products is the MEROPS database ([13] and www.merops.ac.uk) (see Chapter 1), which is maintained by the editors of the *Handbook of Proteolytic Enzymes* [14]. This lists proteases by type and by organism, and includes links to a variety of key information sources. A major advantage of starting from the MEROPS is the classification system that it uses — this is one of the most sophisticated systems around and is, for instance, used in the *D. melanogaster* genome annotation. The listing format makes it is very easy to link between organisms when interested in particular classes of protease, and the availability of precomputed sequence alignments is very useful. Table 1 summarizes the information as it currently stands on the different types of protease in a variety of model systems, abstracted directly from the MEROPS database, and shows some of the significant differences in distribution that can be found between species.

The data reflect the different ways in which proteases have been identified, rather than being a definitive set (this is of course unavoidable, when reports of proteases can be made based on protein sequence data, nucleic acid predictions, or biochemical activity only). The consequence is that any one protease may be represented several times in the lists, until definitive identification has taken place, and so, for some organisms, the groups may be larger than expected. This kind of redundancy should, in principle, disappear for organisms whose genomes have been completely sequenced but, as mentioned above, gene prediction algorithms are not perfect and only functional expression and/or full genetic mapping can confirm that a previously recognized protein or biochemical activity is the same as a predicted gene. This of course works the other way, as we can see from the matrix metalloprotease example that is given later in the chapter.

## Genome databases

The presentation of information for the different model organisms whose genomes have been sequenced has been highly variable, and depends to some extent on the underlying database structure, and on the interface used. For *C. elegans*, the underlying database is ACeDB (**A** *C. elegans* **D**ata**B**ase; www.acedb.org), originally written to hold sequence information and the annotations that go with it. A series of additional programs could then be used to visualize the information for different purposes. The data are all precomputed and therefore require updates. While the sequencing project was in full flow, this was not a problem, but now that the main sequencing effort is essentially finished, the database risks suffering from what will be a common problem for such centralized collections of data, the cost in time and money of curation. A second feature of the database, which is now being addressed much more proactively, is derived from the need to run the software locally. The original version of ACeDB was written for Unix computer systems. The need for local copies led to the creation of a variety of ported versions, for Mac and PC, but did not get rid of the need of having a local 'manager' to look after it. At present, a highly user-friendly and sophisticated web-based interface, WormBase, has been developed (see [15] and www.wormbase.org), which addresses the access problems, and makes finding and working with data much easier. This apparent difference between the production of highly sophisticated display and analytical software, and the difficulty of sharing it outside the specialist user community reflects the changes in computer and networking technology over the period of this seminal sequencing project.

The development of distributed data systems means that information is not held by a single database, such as ACeDB, and disseminated to the users over the net, but that different kinds of data are held in different kinds of databases in different places and are linked together by an overarching interface (a 'portal'). SRS, the sequence retrieval system (www.lionbio.co.uk) and the annotation of the *D. melanogaster* genome linked through to the genetic data of Flybase (http://www.flybase.org/) are good examples of such a distributed database interface. Curation becomes more complex, because each 'section' of the data needs to be looked after; distinct parts of the data can be managed by different groups, as long as there is a Steering Committee to monitor the presentation and the content, but maintenance itself becomes a distributed resource, and hence easier to garner.

Both models use, mainly, precomputed data, but the information that they hold can be very different. Because the locally run ACeDB model did not easily link to other databases, changing the data involved rewriting the models that define the data, recomputing the whole set and redistributing the data. This can mean that it is difficult to take advantage of efforts elsewhere on particular gene families or protein products, such as the proteases. Flybase, WormBase, and the new projects that are now online are structured to allow links directly to specialist resources and databases, by acting as portals, which gives much more flexibility to the researcher seeking answers.

As I suggested above, the data presented in this way will still be the result, at some stage, of interpretation — by the computer algorithms used, as well as by the human annotators. Not only are the predictions not necessarily completely right, they may also be completely wrong and/or incomplete. For example, when the original *D. melanogaster* sequence datasets were analysed, two gene prediction programs were used to analyse the data. In a previous distributed experiment, groups were asked to annotate a large segment of the X chromosome using their favourite analytical techniques [16]. The results from this survey were important in the decision to use Genie (www.fruitfly.org/seq_tools/genie.html — see [17]) and Genscan (http://genes.mit.edu/GENSCAN.html and [18]) as the core prediction program, with Genie the preferred method, because it had been optimized for *D. melanogaster* genes. These use different approaches to gene prediction, and should therefore complement each other. These computations give three sets of predicted genes: those predicted by both programs (obviously a very good bet), those predicted only by Genie, and those predicted only by Genscan. The second group, by Genie, were considered better 'bets' than the third group, though all three sets were included. How comprehensive the identification of genes was is difficult to assess, and there is still some debate as to how big the real gene set is, not only in *D. melanogaster* (see [19]), but to some extent in *C. elegans*. The problem is not only one of identifying all possible gene sequences, but also the relevance of those gene sequences: are the predicted genes expressed or are they pseudogenes (either transcriptionally or translationally silent)? The total set of a protease gene family can be relatively simple to identify, because levels of sequence similarity are high; however, determining whether they are all expressed, or whether some are pseudogenes, requires functional and expression data, which means that any gene prediction that is not associated definitively with a cDNA or a phenotype has to be confirmed independently by identifying the cDNA (often using PCR), or by a full genetic analysis.

### M10 and M2 proteases

Some simple examples will illustrate the problem, and how bringing together all the relevant information can help. The initial analysis of the *D. melanogaster* genome identified, among all the protease families, large sets of zinc metalloproteases. A subset of these, including the matrix metalloproteases (see Chapter 3), have a very clear motif (HEXXHXXGXXH in the M10 and M12 families) associated with the active site, but different families have

**Table 1. Peptidase groups of the major sequenced genomes**

Data summarized from the MEROPS database for the yeast *Saccharomyces cerevisiae*, human, *D. melanogaster*, *C. elegans* and cress (*Arabidopsis thaliana*) (January 2002). The clan identifiers are those described within MEROPS (www.merops.ac.uk), with the following alterations: Clans S(PA), C(PB) and T(PB) should properly be written as PA(S), PB(C) and PB(T), but have been clustered here by catalytic domain type, rather than evolutionary origin (see [27]). Bold values are the total values for each catalytic domain type. The M10A [Clan MA(M)] and M2 [Clan MA(E)] are included for comparative detail.

| Type | Clan | Yeast | Man | Fly | Worm | Cress |
|---|---|---|---|---|---|---|
| Aspartic | AA | 11 | 16 | 37 | 23 | 107 |
| | AX | | 2 | 1 | 3 | 2 |
| | | **11** | **18** | **38** | **26** | **109** |
| Cysteine | C(PB) | 2 | 3 | 3 | 2 | 4 |
| | CA | 19 | 80 | 35 | 61 | 79 |
| | CD | 2 | 20 | 10 | 9 | 7 |
| | CE | 2 | 6 | 4 | 6 | 7 |
| | CF | | 1 | | 1 | 2 |
| | CH | | 3 | 2 | 10 | |
| | CJ | | 3 | 2 | 2 | 6 |
| | CK | | 1 | 1 | | 4 |
| | CX | 1 | 4 | 2 | 2 | 3 |
| | | **26** | **121** | **59** | **93** | **112** |
| Metallo | MA(E) | 11 | 34 | 59 | 43 | 27 |
| | M2 | | 24 | 6 | 1 | |
| | MA(M) | | 67 | 25 | 57 | 6 |
| | M10 | | 24 | 2 | 6 | 5 |
| | MC | 1 | 16 | 23 | 12 | 3 |
| | ME | 8 | 8 | 7 | 9 | 12 |
| | MF | | 3 | 8 | 2 | 3 |
| | MG | 5 | 10 | 9 | 11 | 11 |
| | MH | 12 | 13 | 17 | 12 | 18 |
| | MJ | 1 | 1 | | 1 | 1 |
| | MK | 3 | 2 | 2 | 2 | 3 |
| | MM | | 1 | 1 | 1 | 3 |
| | MX | 1 | 4 | 6 | 1 | |
| | | **42** | **159** | **157** | **151** | **87** |
| Serine | S(PA) | 1 | 106 | 193 | 8 | 12 |
| | SB | 4 | 9 | 4 | 5 | 57 |
| | SC | 5 | 16 | 19 | 31 | 82 |
| | SE | | | | 5 | 2 |
| | SF | 3 | 4 | 3 | 2 | 11 |

**Table 1. (contd.)**

| Type | Clan | Yeast | Man | Fly | Worm | Cress |
|---|---|---|---|---|---|---|
| | SK | | 1 | 1 | 1 | 12 |
| | SM | | | | | 3 |
| | SN | | | 1 | | |
| | SX | 3 | 4 | 4 | 10 | 18 |
| | | **16** | **140** | **225** | **62** | **197** |
| Threonine | T(PB) | 15 | 23 | 34 | 21 | 38 |
| | | **15** | **23** | **34** | **21** | **38** |

particular additional features that define which class they belong to. One consequence of this is that identifying a gene with an HEXXH motif is only the first step in classifying it. Most of the time there is no problem, in that overall sequence similarity, or the use of HMM approaches, will clearly identify which family the predicted product would belong to. On first analysis, it was clear that a whole class of metalloproteases, the matrix metalloproteases (M10), were missing (see Table 1). Closer analysis showed that there was something that looked like an M10 protease, but it lacked some of the key features required for a matrix metalloprotease. However, genetic and biochemical data strongly implied that there was a 'classical' matrix metalloprotease that was identified immunologically and biochemically [20]. As a result of this information, the region of genomic DNA containing the exon with the M10 active site was looked at much more closely and, by playing with the predicted exons in that region and looking for areas of sequence similarity with the cannonical M10 proteases, it was possible to build a gene prediction that matched an M10 matrix metalloprotease very closely, including a propeptide (the 'cysteine switch' typical of matrix metalloproteases) region at the N-terminus, which is coded for in the first exon. Originally called CG4859 in the annotated genome, this prediction has now been confirmed experimentally as the gene Dm1-MMP [21]. A second gene prediction could then be made (CG1794), knowing that the problem of automatic prediction of matrix metalloprotease gene structures existed. At present, there is no other evidence that this is a 'real' gene. The reason why they were not originally predicted is probably a function of the gene prediction programmes — both had large introns, in one case up to 70 kb (Figure 1a). These appear to be the only two M10 matrix metalloproteases in the *D. melanogaster* genome. As this is a relatively large family in humans, and indeed in *C. elegans*, this implies a fundamental functional and developmental difference that awaits further detailed study.

Another example can be seen in the predictions associated with the M2 class, the angiotensin-converting enzyme (ACE)-like proteases (Table 1). ACE has long been known as a key enzyme in the control of blood pressure in vertebrates [22] (see Chapter 10), and its discovery in *D. melanogaster* raised the question of its biological function [23]. A key observation in humans was the duplicated domain structure found in the somatic form of the enzyme, which had been felt to be the result of a relatively recent duplication event. Finding one ACE-like gene (*Ance*) [24] in *D. melanogaster* fits nicely with such a theory; finding two separate copies (*Ance* and *Acer*) [25] made life a little more complicated. As the genome was sequenced, several more copies were found (a total of five), which could have totally confused the picture. However, as with the M10 class, one critical copy was not identified because of a large intron which could be built manually to join two sections, separately identified as having identity with ACE, but which together looked like ACE (Figure 1b). Indeed, there are three predicted genes coded for on the other DNA strand

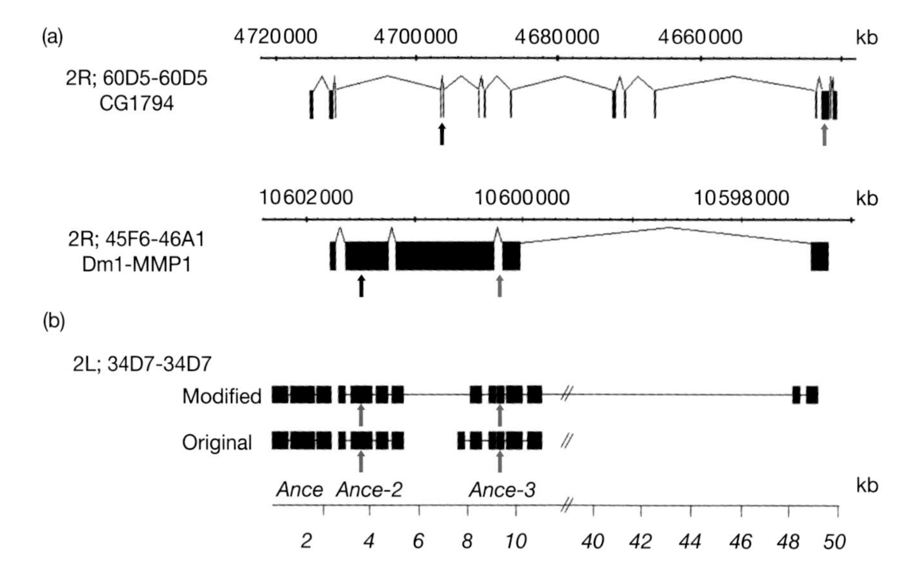


**Figure 1. Comparison of the genome structure of the ancestral duplication of ACE**
(a) Predicted intron–exon structure of the two M10 matrix metalloprotease class genes from *D. melanogaster*. In CG1794, the small exons (represented by vertical black bars) are embedded in large introns, with an overall size of approx. 70 kb. The zinc-binding active site (blue arrow) and the propeptide region (black arrow), which together were used to predict these structures, are indicated. (b) 'Before and after' representations of the *Ance* gene region. Originally, the gene prediction programs suggested the presence of *Ance* (already known), and two further genes whose predicted products had significant partial identity with the Ance protein. The blue arrows mark the sites where the zinc-binding region would be found. Closer examination of the sequence identified a region of approx. 40 kb downstream of *Ance-3*, which not only showed identity with ACE, but included a predicted translation which would put a C-terminal membrane anchor onto the predicted protein, as is found in the vertebrate two-domain ACE gene. Re-analysis of the sequence including these specific regions showed that a duplicated gene structure could be constructed which would code for a double-domain, C-terminal membrane-anchored ACE-like gene.

within this large intron, including one for which Expressed Sequence Tag data exist (BG:DS00180.3). As a result of close analysis of these predicted gene structures, it would appear the duplication found in humans is at least as ancient as the split between humans and flies, and that the multiple copies in *D. melanogaster* are later events [26].

## Conclusion

To go back to my point at the start: while the sequencing and gross annotations are superbly impressive, they are still only the beginning of the task — detailed analysis of the gene families in functional and evolutionary terms will be how the fact of translation from information to life is teased out.

### Summary

- *Protease data mining can take advantage of the many specialist Web-available databases that cover genetic, protein and nucleic acid sequence information specific to a variety of organisms.*
- *Precomputed data, such as gene predictions, should be used with care. Unless there is definitive supporting information, either genetic or biochemical, for a gene prediction, it should be treated with a great deal of caution.*
- *Whereas gene predictions supported by other data are now highly accurate, predictions without such supporting data need some form of confirmation, ideally sequencing of a cDNA to show that the predictions are accurate, followed by expression and biochemical characterization of the predicted protein.*
- *The study of proteases benefits from a flexible but defined classification system, important for comparative work, such as the one used in MEROPS.*

### References

1. The *C. elegans* Sequencing Consortium (1998) Genome sequence of the nematode *C. elegans*: a platform for investigating biology. *Science* **282**, 2012–2018
2. Myers, E., Sutton, G.G., Delcher, A.L., Dew, I.M., Fasulo, D.P., Flanigan, M.J., Kravitz, S.A., Mobarry, C.M., Reinert, K.H., Remington, K.A. et al. (2000) A whole-genome assembly of Drosophila. *Science* **287**, 2196–2204
3. Southan, C. (2000) Assessing the protease and protease inhibitor content of the human genome. *J. Pept. Sci.* **6**, 453–458
4. Stein, L. (2001) Genome annotation: from sequence to biology. *Nat. Rev. Genet.* **2**, 493–503
5. Eddy, S.R. (1998) Profile hidden Markov models. *Bioinformatics* **14**, 755–763
6. Altschul, S.F., Gish, W., Miller, W., Myers, E.W. & Lipman, D.J. (1990) Basic local alignment search tool. *J. Mol. Biol.* **215**, 403–410
7. Pearson, W.R. (1990) Rapid and sensitive sequence comparison with Fastp and Fasta. *Methods Enzymol.* **183**, 63–98

8. Altschul, S.F., Madden, T.L., Schaffer, A.A., Zhang, J., Zhang, Z., Miller, W. & Lipman, D.J. (1997) Gapped BLAST and PSI-BLAST: a new generation of protein database search programs. *Nucleic Acids Res.* **25**, 3389–3402
9. Rubin, G.M., Yandell, M.D., Wortman, J.R., Gabor Miklos, G.L., Nelson, C.R., Hariharan, I.K., Fortini, M.E., Li, P.W., Apweiler, R., Fleischmann, W. et al. (2000) Comparative genomics of the eukaryotes. *Science* **287**, 2204–2215
10. Apweiler, R., Attwood, T.K., Bairoch, A., Bateman, A., Birney, E., Biswas, M., Bucher, P., Cerutti, L., Corpet, F., Croning, M.D. et al. (2000) InterPro – an integrated documentation resource for protein families, domains and functional sites. *Bioinformatics* **16**, 1145–1150
11. Hubbard, T., Barker, D., Birney, E., Cameron, G., Chen, Y., Clark, L., Cox, T., Cuff, J., Curwen, V., Down, T. et al. (2002) The Ensembl genome database project. *Nucleic Acids Res.* **30**, 38–41
12. The Gene Ontology Consortium (2001) Creating the gene ontology resource: design and implementation. *Genome Res.* **11**, 1425–1433
13. Rawlings, N.D. & Barrett, A.J. (2000) MEROPS: the peptidase database. *Nucleic Acids Res.* **28**, 323–325
14. Barrett, A.J., Rawlings, N.D. & Woessner, F.W. (eds) (1998) *Handbook of Proteolytic Enzymes*. Academic Press, London
15. Stein, L., Sternberg, P., Durbin, R., Thierry-Mieg, J. & Spieth, J. (2001) WormBase: network access to the genome and biology of Caenorhabditis elegans. *Nucleic Acids Res.* **29**, 82–86
16. Reese, M.G., Hartzell, G., Harris, N.L., Ohler, U., Abril, J.F. & Lewis, S.E. (2000) Genome annotation assessment in Drosophila melanogaster. *Genome Res* **10**, 483–501
17. Reese, M.G., Kulp, D., Tammana, H. & Haussler, D. (2000) Genie – gene finding in Drosophila melanogaster. *Genome Res.* **10**, 529–538
18. Burge, C.B. & Karlin, S. (1998) Finding the genes in genomic DNA. *Curr. Opin. Struct. Biol.* **8**, 346–354
19. Andrews, J., Bouffard, G.G., Cheadle, C., Lu, J.N., Becker, K.G. & Oliver, B. (2000) Gene discovery using computational and microarray analysis of transcription in the Drosophila melanogaster testis. *Genome Res.* **10**, 2030–2043
20. Woodhouse, E., Hersperger, E., Stetler-Stevenson, W.G., Liotta, L.A. & Shearn, A. (1994) Increased type-IV collagenase in lgl-induced invasive tumors of Drosophila. *Cell Growth Differ.* **5**, 151–159
21. Llano, E., Pendas, A.M., Aza-Blanc, P., Kornberg, T.B. & Lopez-Otin, C. (2000) Dm1-MMP, a matrix metalloproteinase from Drosophila with a potential role in extracellular matrix remodeling during neural development. *J. Biol. Chem.* **275**, 35978–35985
22. Corvol, P., Williams, T.A. & Soubrier, F. (1995) Peptidyl dipeptidase A - angiotensin-I-converting-enzyme. *Methods Enzymol.* **248**, 283–305
23. Isaac, R.E., Coates, D., Williams, T.A. & Schoofs, L. (1998) Insect angiotensin-converting enzyme: comparative biochemistry and evolution. In *Recent Advances in Arthropod Endocrinology* (Coast, G.M. and Webster, S.G., eds), pp. 357–378, Cambridge University Press, Cambridge
24. Cornell, M.J., Williams, T.A., Lamango, N.S., Coates, D., Corvol, P., Soubrier, F., Hoheisel, J., Lehrach, H. & Isaac, R.E. (1995) Cloning and expression of an evolutionary conserved single-domain angiotensin-converting enzyme from Drosophila melanogaster. *J. Biol. Chem.* **270**, 13613–13619
25. Taylor, C.A.M., Coates, D. & Shirras, A.D. (1996) The Acer gene of Drosophila codes for an angiotensin-converting enzyme homologue. *Gene* **181**, 191–197
26. Coates, D., Isaac, R.E., Cotton, J., Siviter, R., Williams, T.A., Shirras, A., Corvol, P. & Dive, V. (2000) Functional conservation of the active sites of human and Drosophila angiotensin I-converting enzyme. *Biochemistry* **39**, 8963–8969
27. Bazan, J.F. & Fletterick, R.J. (1988) Viral cysteine proteases are homologous to the trypsin-like family of serine proteases - structural and functional implications. *Proc. Natl. Acad. Sci. U.S.A.* **85**, 7872–7876

# Subject index

## A



## B



## C

## R



## S

## T



## U



## V



## X



## Z